PERTURBATION THEORY AND THE NUCLEAR MANY BODY PROBLEM

PERTURBATION THEORY AND THE NUCLEAR MANY BODY PROBLEM

Kailash Kumar

DOVER PUBLICATIONS, INC.
MINEOLA, NEW YORK

Bibliographical Note

This Dover edition, first published in 2017, is an unabridged republication of the work originally published in 1962 by North-Holland Publishing Company, Amsterdam, and Interscience Publishers, Inc., New York.

Library of Congress Cataloging-in-Publication Data

Names: Kumar, Kailash, author.
Title: Perturbation theory and the nuclear many body problems / Kailash Kumar.
Description: Mineola, New York : Dover Publications, Inc., [2017] | "This Dover edition, first published in 2017, is an unabridged republication of the work originally published in 1962 by North-Holland Publishing Company, Amsterdam, and Interscience Publishers, Inc., New York"--Title page verso. | Includes bibliographical references and index.
Identifiers: LCCN 2017020408 | ISBN 9780486818955 | ISBN 0486818950
Subjects: LCSH: Perturbation (Quantum dynamics) | Many-body problem.
Classification: LCC QC174.5 .K8 2017 | DDC 530.14/3--dc23
LC record available at https://lccn.loc.gov/2017020408

Manufactured in the United States by LSC Communications
81895001 2017
www.doverpublications.com

PREFACE

For me some of the most difficult passages of the literature of physics have been those which start with the phrases 'it is well known' or 'it can be easily shown'. It has often turned out that 'well known' actually means 'well known in a certain circle', and 'easily shown' really stands for 'easily shown if you have mastered the contents of some previous papers – quoted herein or not'. Such attitudes are perhaps unavoidable in an esoteric discipline like ours. In this book an attempt has been made to qualify most of the significant statements by indicating a reference where justification may be found. The referencing is still not so complete as it possibly could be – especially for periods before 1950 and after mid-1961. I have to apologise to those whose work may have been overlooked. For the book itself, I make no claims to originality, unless in some way the whole be greater than the sum of its parts.

This book arose out of a series of lectures given both at the Tata Institute of Fundamental Research in Bombay and at the Australian National University. A set of notes cyclostyled at the A.N.U. was circulated to some physicists. Perhaps it will be kinder not to mention by name many of those to whom I am indebted for comments on the lecture notes and for correspondence on the subject. I am especially thankful to Dr. D. C. Peaslee for carefully reading through the entire manuscript and making many useful suggestions on the manner of presentation and questions of grammar. To Professor K. J. Le Couteur I am greatly indebted for constant encouragement and help – not only in making available the excellent facilities of his department but also in providing an introduction to the publishers and some valuable comments on parts of the book. I would also like to thank Mr. R. K. Bhaduri for his enthusiastic help in preparation of the first draft in Bombay, and Mrs. Nerdal and Mrs. Edmonds for secretarial help.

During the preparation of the book I have enjoyed a Visiting Fellowship at the Australian National University while on leave from the Tata Institute of Fundamental Research.

Canberra, 1962 K. K.

CONTENTS

III. REARRANGEMENT METHODS: REACTION MATRIX

IV. METHODS OF SOLVING t-MATRIX EQUATIONS AND APPLICATION TO NUCLEAR PROBLEMS

V. CONNECTION WITH SOME OTHER METHODS

GENERAL INTRODUCTION AND PLAN

In the past few years the many body problem in the non-relativistic quantum mechanics has been discussed from many points of view and approximations suited to different physical situations have been developed. At the present moment there does not exist any single framework in which all these ideas may be fitted. Perhaps the most important stimulus to the recent developments has come from perturbation theory [1] which also provides a comprehensive framework for the discussion of a large number of the existing theories. The main approaches can be well exemplified by considering the development of the nuclear many body problem and a fairly complete story can thus be made concerning the topics mentioned in the title.

However, in my view the situation is not quite so satisfactory as has often seemed from the first reading of many papers on the subject. What we have available is only a set of ideas and techniques which lead to results apparently having the right form. Questions of convergence are as intractable here as anywhere else. Actually the situation is far worse than in the case of quantum electrodynamics. In quantum electrodynamics there exists a unique way of separating the divergences, and after this has been done the resulting series is an expansion in powers of a small quantity. Strictly speaking, of course, even this is not any guarantee for convergence. In the non-rela-

[1] For a rapid general review of other methods see the book by ter Haar (t.H1). A fairly complete bibliography will also be found there. Of the more recent developments we refer in particular to the canonical transformation methods of Bogoliubov and others (B1, B2); the theories based on transformations intended to bring out the collective modes (B3 to B6, T1) and the non-canonical non-unitary transformations used in connection with the strong coupling meson theory (P1). All these, their modifications and combinations, may have some utility for the general eigenvalue problem of many interacting particles. Quite often these theories suffer from the defect that an estimation of error is not possible even in principle. This difficulty may be partly alleviated by relating them to variational principles or by striking some comparison with perturbation methods.

tivistic many body problem, however, the infinities have to be, as it were, patched up and the resulting expressions are not expansions in powers of any small parameter. Often in the literature one comes across certain discussions which concern the numerical magnitudes of the first few terms and are supposed to be 'qualitative' arguments concerning the convergence. Also one comes across certain allusions to the connection between different approaches. Many such things are vague and in a larger context unhelpful. No doubt this state of affairs has resulted partly from the fact that the subject is difficult and the literature has grown very rapidly. Therefore I have felt that in writing a book about the subject at this time a different style should be adopted. Thus, vague so-called 'qualitative', discussions concerning the connection between different methods and their validity have been deliberately left out. On the other hand, some pains have been taken to explain the meaning of the highly picturesque and often mystifying language of the modern literature. Such an approach strips the subject of its glamour and gives a certain brittleness to the narrative. It might also cause a feeling of discomfort but it is hoped that one will in that way be closer to an appreciation of the actual situation. Many of the papers on these subjects have been superseded and even the best papers contain some statements and results which are vague or incorrect. This makes it very difficult and equally unprofitable to provide a complete critical evaluation of the literature. However, it is hoped that the following will provide a fairly complete and critical *statement* of the present situation in the field.

The book is primarily meant as an introduction to those who wish to specialise. We have not tried to enforce a complete uniformity of notation because often different notations are convenient for different purposes and also because this will facilitate the transition to journal literature. For instance, when one comes to the non-zero temperature case [2] one finds a large number of different types of diagrams and conventions. It is desirable that one should get used to such practices. We have, therefore, simultaneously used two types of diagrams in the text. Although it does give the figures an appearance of complexity it will be seen that it does not make things any more difficult.

[2] A list of references may be found in (M12), see also (B7, B37, F1, K1, L1, M13). However, our chief concern in the following will be with the systems at absolute zero of temperature.

Some portions of text dealing with more difficult aspects of theory or with questions of detail have been put in small print. A list of main symbols has been appended to each chapter for easy reference.

After this explanation of the general point of view we turn to a brief description of the plan of the book.

It was considered appropriate to start in Chapter I with a discussion of the basic ideas of the perturbation theory. What are the conditions under which the perturbation theory may be set up and what are the various forms of the perturbation expansions?

The problem of existence and convergence of the series associated with the first part of the question has been answered only for a few cases which are not particularly helpful in the many body problem. We can do no more than call attention to this important question. The connection between different forms of expansions is well brought out by following the method of Riesenfeld and Watson (R1) where different choices of a certain operator lead to different expansions. One assumes that these series exist and are meaningful and then tries to compare the relative rates of decrease of the first few terms in order to decide which series to prefer. Also in the same spirit one compares the results of a perturbation calculation to a given order with that of a variational principle suggested by it or by certain freedoms associated with the splitting of the total Hamiltonian in an unperturbed part and a perturbation. If it were possible to carry any of these methods to the limit of the order $n \to \infty$, one would know about the main question of convergence also. Apart from this the methods have an interest of their own as techniques for evaluation of effects up to finite orders; since it is to be hoped that after the questions of convergence are settled, the forms of the final expressions will be similar to those from finite order calculations. To a large extent the specific nature of the interaction does not play any part in these discussions but in the last section we introduce two-body interactions to show how new properties are seen in the expansions on introducing these details. The methods used in the first chapter are purely algebraic and it will be seen that even though the diagrammatic methods of later chapters provide improvements in classification of terms the basic questions of convergence are still very similar.

Further progress in many body perturbation theory has depended upon diagrammatic methods, which provide the most convenient method of classifying the terms. The formal expressions for energies and wave functions can be greatly simplified and it becomes possible to discuss the dependence of various quantities on the volume or number of particles in the system. Various rearrangement methods which depend on the selection and summation of a class of terms having certain desirable characteristics are also greatly facilitated and often suggested by diagrammatic representation of the terms. Therefore, we have considered these topics in detail in Chapters II and III. In Chapter II we start by showing how a simple reduction of the terms of the Rayleigh-Schrödinger series by means of determinantal (Fermion) wave functions into two-particle matrix elements gives rise to the idea of the diagrams. This straightforward approach serves to connect the results of later second quantisation methods to those of Chapter I. At the same time it helps us to make some points concerning some exceptional situations. The rest of this chapter is concerned with a detailed description of the techniques used in second quantisation methods. Expressions for energies and wave function are derived and the important general theorems regarding the systems are proved. Practically none of the results are universally true and about the mathematical rigour one can say very little indeed. We have tried to indicate in the footnotes as much of the limitations as are known to us. It is hoped that this expedient will suffice till we have the final version of the theory. The discussion of rearrangement methods in Chapter III follows the same pattern. We look at these both from the purely algebraic and from the diagrammatic points of view. The case we consider is, of course, the one most useful in the nuclear studies, viz., the t-matrix and associated propagator modifications, but it is obvious that other rearrangements are also possible, and in other problems they may be more strikingly important.

Chapter IV is concerned mainly with techniques of solving the t-matrix equation and other equations that arise in the nuclear many body problem. All methods described in this chapter are of necessity approximate. The intuitive reasoning behind each of them is given, and in the systematic development it is easy to see how they are connected. But the question of their relative merits or their validity can not be easily discussed. We have given references to papers in which such discussions are given but, as mentioned earlier, in the very nature of things, any discussion of the magnitudes of a few terms can not be very helpful. We have further elaborated upon this remark at appropriate places in the text and especially in the section on the accuracy of calculations (Ch. IV, Sec. 6d). For the same reason and because of the ever changing parameters of the two-body nuclear potential we have not tried to give any numerical results but the well-established qualitative features of our understanding have been stated. It is often said that this theory has provided a justification for the shell model. A more accurate statement would be that it has provided a scheme according to which calculations of nuclear properties may be performed with finite, well behaved two-body potentials. It gives, in principle, a prescription for finding these effective potentials from the actual two-body potential which need not be well behaved and, in particular, shows that the treatment of 'configuration-mixing' should be different from that according to earlier ideas but fortunately the same as it was in earlier practice. This topic is dealt with in the last section of Chapter IV.

In Chapter V we have collected various other methods of approaching the many body problem and show how the results may be compared with those of previous chapters. Many of these connections, in particular that with the theory of superconductivity, are not fully worked out. However, it seems that a way has been found to circumvent the most glaring inconsistencies. Whether these procedures are fully justified is a different question whose answer will depend on the successful resolution of some of the questions raised in the following, particularly in Chapters I and V.

Note on numbering of equations

In each section equations are numbered anew in the usual way, giving the section number and the equation number. When reference is made to the same section the section number is omitted. When reference is made to a different chapter the number of the chapter in which the equation occurs is prefixed to the usual number.

CHAPTER I

ALGEBRAIC METHODS

1. INTRODUCTION

The Hamiltonian, H, of the total system is split into two parts – the zeroth order Hamiltonian or unperturbed part, K_{op}, and the interaction or perturbation part V.

$$H = K_{op} + V; \quad H\Psi_\lambda = E_\lambda \Psi_\lambda; \quad \lambda = 0, 1, 2, 3 \ldots \quad (1.1)$$

Eigenvalues and eigenfunctions of K_{op} are supposed to be known [3]. For simplicity we shall assume that K_{op} has a non-degenerate spectrum with a finite or a denumerably infinite number of discrete levels. These assumptions are not essential. Transition to a degenerate or a continuous spectrum is easily made.

Let

$$K_{op}\Phi_l = K_l \Phi_l; \quad l = 0, 1, 2, \ldots \quad (1.2)$$

We adopt the convention that $K_0 \leqslant K_l \leqslant K_{l+1}$. In usual practice l and λ are sets of quantum numbers (*e.g.*, $n\,l\,j\,m$ for atomic levels) which depend on the symmetries [4] of K_{op}, V and H.

The purpose of perturbation theory is to calculate the quantities E_λ, and Ψ_λ from the known quantities K_l, Φ_l.

There are as many Ψ_λ's as there are Φ_l's and the main problem of perturbation theory is to establish a one to one correspondence between the two sets $\{\Phi\}$ and $\{\Psi\}$. This is done by specifying for each Φ_l the Ψ_λ that tends to it as V tends to zero. (To be exact put $H = K_{op} + \varepsilon V$ and let $\varepsilon \to 0$.) When such a correspondence is established we will use the same numbers for l and λ. Thus Ψ_0 will be that eigenfunction of H that goes to Φ_0 of K_{op} as V goes to zero.

The Zeeman effect (C1) provides us with an example where the

[3] We depart from the usual notation only in this chapter. In other chapters we shall revert to the usual notation in which H_0 ($= K_{op}$) is the unperturbed Hamiltonian with eigenvalues E rather than K. *E and E_λ for perturbed eigenvalues are used only in this chapter.* See the list of symbols at the end of this and other chapters.

[4] For a group theoretic treatment of this topic see for instance (H8) and (L7).

perturbation may be controlled and varied. In presence of the perturbing field only n, m, m_s are the constants of motion, l (or j) is not. However, we can use the field free set of quantum numbers $n\ l\ m\ m_s$ to denote the levels in the presence of the field also since we know which of the perturbed levels go into which of the unperturbed levels.

More often it is not within our powers to physically change V. We cannot gradually reduce the nuclear interaction to observe the continuous transition from Φ_0 to Ψ_0.

Actually in all cases the connection between the two spectra is to be established mathematically. This 'pairing' of indices of two sets of eigenvalues enables us to use the same quantum numbers for both $\{\Psi\}$ and $\{\Phi\}$, even though they may not be good quantum numbers for the former, and suggests that Ψ_λ may be expressed as a power-series (in ε) involving matrix elements $(\Phi_l, V\Phi_k) = V_{lk}$ of V. The leading term of the expansion would naturally be $\Phi_{l=\lambda}$.

There are cases where there is no clear one to one correspondence between the perturbed and unperturbed spectra. For example, consider a square well potential of depth V_1 such that it has n_1 bound states and the continuum and another similar potential of depth V_2 such that it has $n_2 \neq n_1$ bound states and the continuum. Then one has difficulty in setting up a perturbation theory based on states of V_1, say, since the spectrum (for V_2) undergoes a discontinuous change as the perturbation $\varepsilon(V_2 - V_1)$ attains its full value $(V_2 - V_1)$. Such singular cases are not well understood [5]. We shall assume that such situations do not arise in our case [6].

2. PERTURBATION THEORY AS A SEARCH FOR THE POLES OF THE GREEN'S FUNCTION (R1)

A particularly compact account of the various perturbation theories can be given in this way. Consider the function $W(E)$ of any complex number E defined by

$$W(E) = \sum_\lambda \frac{|\Psi_\lambda\rangle\langle\Psi_\lambda|}{E - E_\lambda}. \tag{2.1}$$

[5] This is an example of non-analytic perturbation. Some related questions have been discussed from a mathematical point of view by F. Rellich (R2).

[6] This is not to say that such cases do not arise in nature, as a matter of fact they occur quite often (see R2). The appearance of superconductivity might be related to a case of this type.

This is known as the Green's function for the system as a whole. It must be distinguished from other one- or more particle Green's functions which are also used in the study of the many body problem. It is clear that the poles of $W(E)$ occur at $E = E_\lambda$.

It is convenient to adopt the bracket notation and work in the so-called 'p' representation in which K_{op} is diagonal. Equations (1.2) and (1) become

$$K_{op}|p\rangle = K(p)|p\rangle \tag{2.2}$$

and

$$\langle p|W(E)|p'\rangle = \sum_\lambda \frac{\Psi_\lambda(p)\Psi_\lambda^*(p')}{E - E_\lambda} \tag{2.3}$$

respectively, where $\Psi_\lambda(p) = \langle p|\Psi_\lambda\rangle$; $\Psi_\lambda = \sum_p \Psi_\lambda(p)|p\rangle$. Using (1) and (1.1)

$$(E - H)W = 1. \tag{2.4}$$

To get actual answers we must express $W(E)$ in terms of quantities known from the solution of the zero order equation (1.2). For this purpose it is convenient to write

$$(E - K_{op})W = 1 + VW. \tag{2.5}$$

Equations (4) and (5) must be considered as operating on a ket $|p\rangle$. This completely defines the problem.

We proceed to separate the diagonal part $W_d(p)$ of the matrix of W in the p-representation and define a quotient matrix $\boldsymbol{M}$.

$$W(E) = \boldsymbol{M}W_d \tag{2.6}$$

$$\langle p|W(E)|p\rangle = \sum_{p'} \langle p|\boldsymbol{M}|p'\rangle \langle p'|W_d|p\rangle \tag{2.7}$$

by definition

$$\langle p|W_d|p'\rangle = \langle p|W|p\rangle\delta_{pp'} = W_d(p)\delta_{pp'} \tag{2.8}$$

hence

$$\langle p|\boldsymbol{M}|p\rangle = 1. \tag{2.9}$$

From (5) and (6) we obtain

$$(E - K_{op})\boldsymbol{M}W_d = 1 + V\boldsymbol{M}W_d. \tag{2.10}$$

Taking the diagonal part:

$$(E - K(p))W_d = 1 + V_c(p)W_d, \tag{2.11}$$

where

$$V_c(p) = \langle p|V_c|p\rangle = \langle p|VM|p\rangle. \tag{2.12}$$

The formal solution of (11) is

$$W_d = [E - K(p) - V_c(p)]^{-1}. \tag{2.13}$$

The poles of W_d occur at the same places as those of W; hence we have the eigenvalues of the perturbed system

$$E_\lambda = E = K(p) + V_c(p). \tag{2.14}$$

The 'energy shift' of the p_0 level of the zero order Hamiltonian is

$$\delta E = E_{\lambda 0} - K(p_0) = V_c(p_0) \tag{2.15}$$

$$\delta E = \lim_{E \to E_{\lambda 0}} \langle p_0|VM|p_0\rangle. \tag{2.16}$$

Multiplying from the right by W_d^{-1} in (10), we have

$$(E - K_{\text{op}})M = W_d^{-1} + VM. \tag{2.17}$$

All these must be considered as operations to the right on some definite eigenfunction of K_{op}, for instance on the ket $|p_0\rangle$. Then as $E \to E_{\lambda 0}$, $W_d^{-1}|p_0\rangle = 0$ and (17) reduces to

$$E_{\lambda 0}M|p_0\rangle = (K_{\text{op}} + V)M|p_0\rangle = HM|p_0\rangle.$$

Hence, the eigenfunction of H is expressed in terms of the matrix M.

$$\Psi_{\lambda 0} = \lim_{E \to E_{\lambda 0}} M|p_0\rangle; \qquad \Psi_{\lambda 0}(p) = \lim_{E \to E_{\lambda 0}} \langle p|M|p_0\rangle. \tag{2.18}$$

The precise manner in which the complex number E tends to the value $E_{\lambda 0}$ at the pole of $W(E)$ determines the boundary conditions satisfied by the wave function $\Psi_{\lambda 0}$. The limiting process will be considered in the next chapter where operators $O_\alpha^\pm$ corresponding to different boundary conditions are defined. For the present it is not necessary to specify the limiting process.

Our problem is now reduced to that of finding the matrix M, which through equations (18) and (16) determines the eigenvalues and eigenfunctions of H and for which an integral equation is obtained by subtracting an operator $\boldsymbol{OM}$ from both sides of (17) and multiplying from the left by $[E - K_{\mathrm{op}} - O]^{-1}$,

$$\boldsymbol{M} = [E - K_{\mathrm{op}} - O]^{-1} W_{\bar{d}}^{-1} + [E - K_{\mathrm{op}} - O]^{-1} [V - O] \boldsymbol{M}. \tag{2.19}$$

This is the basic integral equation for $\boldsymbol{M}$ in which all quantities are, in principle, known (W_d from (13) and (12)). The operator O, which is arbitrary, can be chosen in such a manner as to give desirable properties to the expressions for $\boldsymbol{M}$.

To remove the odd looking inhomogeneous term of (19) let

$$\boldsymbol{M} = MD, \tag{2.20}$$

$$D = [E - K_{\mathrm{op}} - O]^{-1} W_{\bar{d}}^{-1}. \tag{2.21}$$

Then

$$M = 1 + [E - K_{\mathrm{op}} - O]^{-1} [V - O] M. \tag{2.22}$$

If O is chosen such that D is diagonal and if it can be arranged that

$$\langle p_0 | D | p_0 \rangle = \langle p_0 | [E - K_{\mathrm{op}} - O]^{-1} W_{\bar{d}}^{-1} | p_0 \rangle \rightarrow \text{unity} \tag{2.23}$$

in the limit of $E \rightarrow \boldsymbol{E}_{\lambda 0}$, then

$$M = \boldsymbol{M} = 1 + \frac{1}{[E - K_{\mathrm{op}} - O]} [V - O] \boldsymbol{M}. \tag{2.24}$$

These equations determine the $|p_0\rangle$ column of the M or $\boldsymbol{M}$ matrix. The limiting procedures implied in (16) and (18) establish the required one to one correspondence between the spectra of H and K_{op}. Strictly speaking, one presupposes that such a correspondence has been achieved and then finds the expression for the perturbed quantities. This is why the singular cases mentioned in the introduction cannot be treated here.

2a. *Some special cases*

Different perturbation theories result from different choices of the operator O. We consider a few examples. Original references are given with the titles. *We are interested here in the energy shift of a particular level* $|p_0\rangle$.

2a (i). Brillouin-Wigner perturbation theory (B9, W1)

$$O = V_c(p_0)\Lambda_{p_0} \tag{2.25}$$

where $\Lambda_{p_0} = |p_0\rangle\langle p_0|$ is the projection operator onto the state p_0. With this O, equations (13) and (21) imply that (23) is satisfied. Hence $\boldsymbol{M} = M$. Because of equation (9)

$$\begin{aligned} O\boldsymbol{M}|p_0\rangle &= V_c\Lambda_{p_0}\boldsymbol{M}|p_0\rangle = \langle p_0|V\boldsymbol{M}|p_0\rangle|p_0\rangle \\ &= \Lambda_{p_0}V\boldsymbol{M}|p_0\rangle \end{aligned}$$

and (24) becomes [7]

$$\boldsymbol{M} = 1 + \frac{1}{E - K_{\text{op}}}[1 - \Lambda_{p_0}]V\boldsymbol{M}. \tag{2.26}$$

Thus the perturbed energy is given by the solution of

$$E = K(p_0) + \langle p_0|V|p_0\rangle + \langle p_0|V\frac{1}{E - K_{\text{op}}}[1 - \Lambda_{p_0}]V|p_0\rangle$$

$$+ \langle p_0|V\frac{1}{E - K_{\text{op}}}[1 - \Lambda_{p_0}]V\frac{1}{E - K_{\text{op}}}[1 - \Lambda_{p_0}]V|p_0\rangle + \ldots \tag{2.27}$$

The series is cut off at a certain number of terms and the lowest root of the equation gives the energy level $E_{\lambda 0}$. The expression for the wave function can be written in a similar way.

2a (ii). Eden-Francis perturbation theory (E1)

$$O \equiv \Delta \equiv \{\delta E + (E - E_{\lambda 0})[1 - |\Psi_{\lambda 0}(p_0)|^{-2}]\}\mathscr{I}. \tag{2.28}$$

$\mathscr{I}$ is the unit operator, henceforward it will not be explicitly written. Because of the singularity

$$\lim_{E\to E_{\lambda 0}} W_d(p_0) = \frac{|\Psi_{\lambda 0}(p_0)|^2}{E - E_{\lambda 0}}$$

where for degenerate cases a modification is required. We have

$$\langle p_0|D|p_0\rangle = \langle p_0|[E - K_{\text{op}} - \Delta]^{-1}W_{\bar{d}}^{-1}|p_0\rangle$$

and

$$\lim_{E\to E_{\lambda 0}} \langle p_0|D|p_0\rangle = \langle p_0|[E - K(p_0) - \delta E - E + E_{\lambda 0} + W_{\bar{d}}^{-1}]^{-1}W_{\bar{d}}^{-1}|p_0\rangle = 1.$$

[7] For arbitrary operators A and B

$$\frac{1}{A + B\Lambda_{p_0}}[1 - \Lambda_{p_0}] = \frac{1}{A}[1 - \Lambda_{p_0}].$$

Hence $\boldsymbol{M} = M$ and

$$M = 1 + \frac{1}{E - K_{\mathrm{op}} - \Delta} [V - \Delta]M \tag{2.29}$$

where the limit should be remembered.

2a (iii). Rayleigh-Schrödinger perturbation theory

This is the perturbation theory we are taught in the first courses on quantum mechanics. Here it appears as a special case related to the Eden-Francis method, or

$$O = \Delta = \delta E + \Lambda_{p_0}(V - \delta E)\Lambda_{p_0} = \delta E + (\langle p_0|V|p_0\rangle - \delta E)\Lambda_{p_0} \tag{2.30}$$

$$\boldsymbol{M} = 1 + \frac{1}{K(p_0) - K_{\mathrm{op}}} [1 - \Lambda_{p_0}][V - \delta E]\boldsymbol{M}. \tag{2.31}$$

δE occurs in the numerator and is equal to $\langle p_0| V\boldsymbol{M}|p_0\rangle$. In contrast to the B–W expansion (26) the expansion here is linear in E.

2a (iv). Tanaka-Fukuda perturbation theory (T2, F2)

$$O = O_{\mathrm{TF}} = \sum_p \delta E_p \Lambda_p \tag{2.32}$$

where Λ_p is the projection operator onto the state $|p\rangle$ and δE_p is the energy shift associated with this level. Since W_d is diagonal

$$\langle p_0|D|p_0\rangle = \langle p_0|[E - K_{\mathrm{op}} - \sum_p \delta E_p \Lambda_p]^{-1} W_{\bar{d}}^{-1}|p_0\rangle \to 1 \text{ as } E \to E_{\lambda 0}.$$

Hence

$$\boldsymbol{M} = M = 1 + \frac{1}{E - K_{\mathrm{op}} - O_{\mathrm{TF}}} [V - O_{\mathrm{TF}}]M. \tag{2.33}$$

Historically the above methods arose in attempts to formulate the perturbation theory in different ways with the hope of finding more useful series. The Eden-Francis (E1) and Tanaka-Fukuda (T2, F2) methods arose mainly in connection with the many body problem.

The criteria for the choice of O are easy to see. First of all it should be such that successive iterations of M or $\boldsymbol{M}$ converge with sufficient rapidity. Then the operator $[V - O]$ must be a quantity which can be handled easily in actual calculations.

For an N particle system interacting through two-body forces E, K, $(E - K) \simeq \delta E$ are all $\mathrm{O}(N)$ compared to the single particle energies, while V is $O(N^2)$. This makes equations (27) and (31) quite useless. The choices (28) and (32) also do not improve the situation very much. One now has to fall back on some further formal manipulations. One way of doing this, the so called 'optical model' methods of Watson (R1), will be discussed later. At the moment let us turn to a more detailed discussion of some properties of the expansions (27) and (31).

We shall discuss some of the variation-perturbation methods which have the B–W and R–S series as their starting points. The modifications to the B–W perturbation theory retain the property of the original method in that an implicit equation for the energy is to be solved at some stage. The modifications to the R–S method retain the property that the perturbed energy is obtained explicitly as a function of unperturbed quantities. Connections between the various methods will also be commented upon.

3. MODIFICATIONS TO BRILLOUIN-WIGNER PERTURBATION THEORY

We represent the operator $[1 - \Lambda_{p_0}]$ by putting a prime in its place. Cutting off the series after $n+1$ terms, we obtain the nth approximant to M of equation (2.26)

$$M^{(n)} = 1 + \frac{1}{E - K_{\text{op}}}\,{}'V + \frac{1}{E - K_{\text{op}}}\,{}'V\,\frac{1}{E - K_{\text{op}}}\,{}'V + \ldots$$

$$+ \underbrace{\frac{1}{E - K_{\text{op}}}\,{}'V\,\frac{1}{E - K_{\text{op}}}\,{}'V \ldots {}'V\,\frac{1}{E - K_{\text{op}}}\,{}'V}_{n\text{- factors}}. \tag{3.1}$$

The approximant to the wave function is written in greater detail as

$$\Psi^{(n)}_{\lambda 0} = \lim_{E \to E_{\lambda 0}} M^{(n)}|p_0\rangle = |p_0\rangle$$

$$+ \sum_{a \neq 0} \frac{V_{a0}}{E - K_a}\,|p_a\rangle + \sum_{a,b \neq 0} \frac{V_{ba}V_{a0}}{(E - K_b)(E - K_a)}\,|p_b\rangle$$

$$+ \ldots + \sum_{u,v,\ldots,a \neq 0} \frac{V_{uv} \ldots V_{ba}V_{a0}}{(E - K_u)\ldots(E - K_a)}\,|p_u\rangle \tag{3.2}$$

where K_a is the eigenvalue of K_{op} corresponding to $|p_a\rangle$ and $V_{ab} = \langle p_a|V|p_b\rangle$. There are n- factors V_{ab} in the last term.

Equation (2.27) gives up to the same order the familiar form

$$E = K(p_0) + V_{00} + \sum_{a \neq 0} \frac{V_{0a}V_{a0}}{(E - K_a)} + \sum_{a,b \neq 0} \frac{V_{0b}V_{ba}V_{a0}}{(E - K_b)(E - K_a)} + \dots$$

$$\dots + \sum_{u,v,\dots,a \neq 0} \frac{V_{0u}V_{uv} \dots V_{a0}}{\underbrace{(E - K_u)(E - K_v) \dots (E - K_a)}_{n-1 \text{ factors}}}. \tag{3.3}$$

This may be solved for E where the lowest root will correspond to the perturbed energy. Note that this is not an expansion in the powers of the interaction parameter $\lambda(V \propto \lambda)$ because of the presence of E in the denominators. A series in powers of λ can be obtained by expanding the denominators also – this would lead to R–S series, which therefore will have in general a smaller radius of convergence. The overall (formal) convergence requires some conditions on the behaviour of the perturbing potential V (M1). However, we are interested in devising a practical method of computation and hence the convergence only in principle is too broad a requirement to be really useful. The following discussion takes such convergence for granted [8] and seeks to find a method which gives a reasonable approximation in a manageable number of steps.

In a series of papers Feenberg (F3, F4, F5, G2, B10, Y1) and his collaborators have discussed modifications to the B–W perturbation theory which bring about some such improvements. These modifications should be studied as examples of attempts to find properties of the equations and the series which are not evident from the integral equations or the power series.

3a. *Variation-iteration procedure* (G2)

Substituting the nth approximant $\Psi_{\lambda 0}^{(n)}$ to the wave function in the expression

$$E = \frac{\langle \Psi_{\lambda 0}^{(n)} | H | \Psi_{\lambda 0}^{(n)} \rangle}{\langle \Psi_{\lambda 0}^{(n)} | \Psi_{\lambda 0}^{(n)} \rangle} \tag{3.4}$$

[8] A particularly instructive example which illustrates the pitfalls of such an approach was pointed out by Wigner (W4) and further discussed by Trees (T3). See also Epstein (E2).

we have the corresponding energy expression

$$E = K(p_0) + V_{00} + \epsilon_2 + \epsilon_3 + \dots + \epsilon_{2n+1}. \tag{3.5}$$

The first few terms on the right hand side are exactly those in equation (3). Feenberg's modification consists in introducing variational parameters [9] G_i, $1 \leqslant i \leqslant n$, in the approximate wave function. Then dropping the subscript λ_0, ($E \equiv E_{\lambda 0}$ below),

$$|\Psi^{(n)}\rangle = |p_0\rangle + G_1 \sum_{a \neq 0} \frac{V_{a0}}{(E - K_a)} |p_a\rangle + $$

$$+ G_2 \sum_{a,b \neq 0} \frac{V_{ba}V_{a0}}{(E - K_b)(E - K_a)} |p_b\rangle + \dots . \tag{3.6}$$

Or

$$|\Psi^{(n)}\rangle = |p_0\rangle + \sum_{k \neq 0} |p_k\rangle [\sum_{i=1}^{n} G_i M_k(i)] \tag{3.6'}$$

where

$$M_k(i) = \langle p_k| \left(\frac{1}{E - K_{\text{op}}} 'V \right)^i |p_0\rangle.$$

With this wave function Eq. (4) gives

$$\begin{aligned} E = K(p_0) &+ V_{00} + (2G_1 - G_1^2)\epsilon_2 + (G_1^2 + 2G_2 - 2G_1G_2)\epsilon_3 \\ &+ (2G_3 - G_2^2 + 2G_1G_2 - 2G_1G_3)\epsilon_4 \\ &+ (2G_2^2 + 2G_4 + 2G_1G_3 - 2G_1G_4 - 2G_2G_3)\epsilon_5 + \dots \\ &+ (2G_nG_{n-1} - G_n^2)\epsilon_{2n} + G_n^2\epsilon_{2n+1}. \end{aligned} \tag{3.7}$$

One obtains the minimum of this expression if

$$\frac{\partial E}{\partial G_i} = 0, \qquad i = 1, \dots n. \tag{3.8}$$

This gives n- linear equations

$$\epsilon_j = \sum_{i=1}^{n} G_i(\epsilon_{i+j-1} - \epsilon_{i+j}), \qquad j = 2, 3 \dots, n + 1. \tag{3.9}$$

[9] If the interaction, V, may be written as a linear combination of terms of different types *e.g.*, different exchange forces, central and tensor forces, $V = \Sigma_x V_x$, then it is more profitable to introduce independent variational parameters $G_i(x)$ corresponding to different potential types (F4).

To find the optimum G_i, define

$$G_i = 1 + g_i, \tag{3.10}$$

so that (9) becomes

$$\epsilon_{j+n} = \sum_{i=1}^{n} g_i(\epsilon_{i+j-1} - \epsilon_{i+j}). \tag{3.11}$$

The solutions of these equations can be written down in terms of determinants involving ϵ's, and the energy expressed as a continued fraction whose approximants are invariant under the scale transformations of section 3b (F3, Y1).

We consider some simple cases:
let $G_i = 0$, $i \neq 1$, then $G_1 = (1 - \epsilon_3/\epsilon_2)^{-1}$ and

$$E = K_0 + \frac{\epsilon_2}{1 - \epsilon_3/\epsilon_2}. \tag{3.12}$$

(Note change of energy origin: put $V_{00} = 0$ on right hand side or redefine $K_0 = K(p_0) + V_{00}$.)

We have to discuss the roots of this equation for E. Only the lowest root is of interest. This falls below K_0 and the lowest root of $\epsilon_2 - \epsilon_3 = 0$. From the form of ϵ's, $\epsilon_2 - \epsilon_3 < 0$ below the lowest root of $\epsilon_2 - \epsilon_3 = 0$. Hence

$$\epsilon_2 + \epsilon_3 > \frac{\epsilon_2}{1 - \epsilon_3/\epsilon_2}$$

for a range of E including the lowest root. This shows that (12) gives a better approximation than the simple expression $E = K_0 + \epsilon_2 + \epsilon_3$.

Next let $G_i = 0$ for $i \neq 2$ or 1 in equation (7). The stationary value is obtained by completing the squares.

$$E = K_0 + \frac{\epsilon_2}{1 - \epsilon_3/\epsilon_2} + \frac{[\epsilon_2\epsilon_4 - \epsilon_3]^2/\epsilon_2}{(1 - \epsilon_3/\epsilon_2)[(\epsilon_2 - \epsilon_3)(\epsilon_4 - \epsilon_5) - (\epsilon_3 - \epsilon_4)^2]}. \tag{3.13}$$

This is a minimum if

$$\epsilon_2 - \epsilon_3 < 0; \quad (\epsilon_2 - \epsilon_3)(\epsilon_4 - \epsilon_5) - (\epsilon_3 - \epsilon_4)^2 > 0$$

hold for a region including the lowest root of (13). Further simplification results if odd order terms vanish [10]; $\epsilon_3 = \epsilon_5 = 0$. We have then

$$E = K_0 + \epsilon_2 + \frac{\epsilon_4}{1 - \epsilon_4/\epsilon_2}. \tag{3.14}$$

By an argument similar to the one given after Eq. (12) it follows that the lowest root of (14) is a better appoximation to the energy than the lowest root of the equation $E = K_0 + \epsilon_2 + \epsilon_4$ is.

One notices in each of the above examples that the present method improves the wave function and the energy values over the perturbation theory values. However, it needs to be verified for each particular case that it is so. The general conditions under which the present method gives better values are not known. We know that as n tends to infinity all G_i should tend to the value unity. Exactly how this comes about is not clear.

3b. *Properties of B–W series and related variational principles under certain transformations* (*F3, F4, G2, Y1*)

We are free to split the actual Hamiltonian H into whatever parts we consider most convenient. But in practice this freedom is almost non-existent because it is very difficult to find an exactly solvable H_0 appropriate to a given H. However, even with a given H_0 and $V = (H - H_0)$ one may add a certain freedom by allowing transformations that do not upset the symmetry properties of the zero order Hamiltonian. One such transformation is a uniform compression of the unperturbed energy spectrum – the so-called scale transformation, and another is a change of origin of the zero-order spectrum. These transformations introduce an extra parameter into the problem but in the calculations one still uses the wave functions and energies of the original H_0. Sometimes these parameters can be used as variational parameters to obtain a minimum of the perturbed energy. However, their most useful role has been in providing insight into the structure of the perturbation method itself

[10] This is a rather artificial restriction. The situation is realised, for example, when the hydrogen atom problem is formulated in an Einstein hypersphere and the Coulomb potential itself is treated as a perturbation (T3). However, it does give rise to some far-reaching simplifications.

and in providing some more convenient (section 3b (ii)) calculational procedures for the evaluation of the B–W polynomial.

Here it may be mentioned that the renormalization procedure is an example of scale transformation where the scale parameter is determined by the self-energy contributions. Also an 'effective-mass' approximation may be obtained by choosing the parameter so that the perturbed energy vanishes up to some given order.

3b (i). Change of scale of unperturbed spectrum

Let

$$H = K'_{\mathrm{op}} + W = K_{\mathrm{op}} + V \tag{3.15a}$$

where

$$K'_{\mathrm{op}} = K_{\mathrm{op}} + (\mu - 1)(K_{\mathrm{op}} - E') \tag{3.15b}$$

so that

$$(K'_{\mathrm{op}} - E') = \mu(K_{\mathrm{op}} - E') \tag{3.15c}$$

and

$$\begin{aligned} W &= V - (\mu - 1)(K_{\mathrm{op}} - E') \\ &= V - [(\mu - 1)/\mu](K'_{\mathrm{op}} - E'). \end{aligned} \tag{3.15d}$$

Equation (15c) above shows that the difference between the present splitting and the former splitting $H = K_{\mathrm{op}} + V$ is equivalent to a uniform change of scale of the unperturbed spectrum.

Now the energy is

$$E' = K_0 - \sum_{n=0}^{\infty} \epsilon'_{n+2} \tag{3.16}$$

where

$$\epsilon'_{n+2} = \sum_{a,b,\ldots,g\neq 0} \frac{W_{0a}W_{ab}\ldots W_{fg}W_{g0}}{(E' - K'_a)(E' - K'_b)\ldots(E' - K'_g)}. \tag{3.17}$$

From (15)

$$W_{ab} = V_{ab} + (\mu - 1)(E' - K_a)\delta_{ab} \tag{3.18a}$$

$$(E' - K'_a) = \mu(E' - K_a). \tag{3.18b}$$

Substituting the above in (17) one gets

$$\epsilon'_{n+2}(E') = \mu^{-n-1}$$

$$\sum_{a,b,\ldots,f,g\neq 0} \frac{V_{0a}(V_{ab}+(\mu-1)(E'-K_a)\delta_{ab}) \ldots (V_{fg}+(\mu-1)(E'-K_g)\delta_{fg})V_{g0}}{(E'-K_a)(E'-K_b)\ldots(E'-K_g)}$$

$$= \mu^{-n-1} \sum_{s=0} \binom{n}{s} (\mu-1)^s \epsilon_{n+2-s}(E').$$

Note that the total energy is independent of μ; since on changing the summation index from s to $t = n - s$ in the expression above and summing

$$\sum_{n=0}^{\infty} \epsilon'_{n+2}(E') = \sum_{n=0}^{\infty} \sum_{t=0}^{n} \binom{n}{n-t} \left(\frac{\mu-1}{\mu}\right)^{n-t} \mu^{-t-1} \epsilon_{t+2}$$

$$= \sum_{t=0}^{\infty} \epsilon_{t+2}\, \mu^{-t-1} \sum_{n=t}^{\infty} \binom{n}{t} \left(\frac{\mu-1}{\mu}\right)^{n-t}$$

$$= \sum_{t=0}^{\infty} \epsilon_{t+2}\, \mu^{-t-1} \left(1 - \frac{\mu-1}{\mu}\right)^{-t-1}$$

$$= \sum_{t=0}^{\infty} \epsilon_{t+2}(E'). \tag{3.19}$$

It can be verified (G2) by explicit calculation that if the variation-iteration procedure of section 3a is used with the first or second order approximants and the splitting (15) of the Hamiltonian then the minimum obtained is independent of the scale parameter μ. The invariance of the infinite order case with $G_i = 1$ was proved above. These observations led Feenberg to conjecture that in all orders the variation iteration procedure will give an (implicit) equation for E which will be independent of μ.

If the variation-iteration procedure is *not* used, then the energy equation and its lowest root are both dependent on μ. This dependence may be further exploited by minimising the energy with respect to μ, thus giving another variational principle.

Feenberg and his collaborators (F3, Y1) have also provided a 'proof' of the invariance of the variation-iteration procedure with respect to the scale transformation. The proof is as follows:

The variational wave function corresponding to the original splitting, $H = K_{op} + V$, is given by (6′). From the same expression one gets the variational wave function for the splitting (15) by modifications of $M_k(i)$ analogous to those of ϵ_n.

$$|\Psi'^{(n)}\rangle = |p_0\rangle + \sum_{k\neq 0} [\sum_{i=1}^{n} G'_i M'_k(i)] |p_k\rangle. \tag{3.20}$$

The wave functions

$$|\Psi^{(n)}\rangle = |\Psi'^{(n)}\rangle \tag{3.21a}$$

provided

$$\sum_{i=1}^{n} G'_i M'_k(i) = \sum_{i=1}^{n} G_i M_k(i). \tag{3.21b}$$

Manipulations as in proving (19) show that

$$M'_k(i) = \frac{1}{\mu^i} \sum_{s=0}^{i} \binom{i-1}{s} (\mu - 1)^s M_k(i-s).$$

Therefore (21*b*) holds if

$$G_t = \frac{1}{(\mu - 1)^t} \sum_{i=1}^{n} \binom{i-1}{i-t} \left[\frac{\mu - 1}{\mu}\right]^i G'_i. \tag{3.22a}$$

If instead one had started with splitting $H = K'_{op} + W$ and then gone to the splitting $H = K_{op} + V$ through the scale transformation with $\mu' = 1/\mu$ then the requirement would have been

$$G'_t = \frac{1}{(\mu' - 1)^t} \sum_{i=1}^{n} \binom{i-1}{i-t} \left[\frac{\mu' - 1}{\mu'}\right]^i G_i. \tag{3.22b}$$

By substituting one in the other it may be verified that an identity results so that the relations are consistent.

The wave functions are equal if (22a, b) are satisfied. In that case numerically

$$E(G_1 \ldots G_2) = E'(G'_1, G'_2 \ldots G'_n). \tag{3.24}$$

E' is *not* the same function of G'_i as E is of G_i. The minimisation conditions yield in the two cases sets of equations linear in G_i

$$\frac{\partial E}{\partial G_i} = 0; \quad \frac{\partial E'}{\partial G'_i} = 0. \tag{3.24'}$$

These equations are not independent. As a matter of fact they imply each other in virtue of (22). Hence the value at the minimum would be the same whether one minimised E or the scale transformed $E'(\mu)$.

This proof does not seem to be complete. It seems to say that a choice of G_i' can consistently be made such that the two procedures give the same result. It does not follow that it is always necessarily so.

It seems that to be consistent with original B–W perturbation theory the variation-iteration procedure must give the result that as $n \to \infty$ all $G_i \to 1$ at the minimum. At the moment one does not know whether this is really the case. It need not be so if there are some (anomalous) states accessible to the variational procedure but not to the ordinary B–W perturbation theory. This is an important question. The investigation of various invariance properties of the method was partly motivated by the hope that it might throw some light on this question but this has not been possible.

3b (ii). DISPLACEMENT OF THE ORIGIN OF UNPERTURBED SPECTRUM (F3, Y1)

$$H = K'_{\mathrm{op}} + W \tag{3.25a}$$

$$K'_{\mathrm{op}} = K_{\mathrm{op}} + U \tag{3.25b}$$

$$W = V - U. \tag{3.25c}$$

The general term of the B–W series in this case is

$$\epsilon_s(E - K_0, U) = \sum_{m,n,\ldots,y,z\neq 0} \frac{V_{0m}(V_{mn} - U\delta_{mn}) \ldots (V_{yz} - U\delta_{yz})V_{z0}}{(E - K_m - U) \ldots (E - K_z - U)}$$

where the number of indices to be summed over is $s - 1$, and U is a constant which is the parameter of the displacement transformation.

The implicit B–W equation for the energy corresponding to n^{th} order wave function is

$$E = K_0 + \sum_{s=2}^{2n+1} \epsilon_s(E - K_0, U).$$

This is to be solved for E as a function of U and then a value of U may be found which minimises it. The proof of the invariance of the whole series w.r.t. this transformation may be developed in a manner similar to that for invariance under the scale transformation. One particular use of this transformation is in reducing the B–W equation to a simpler polynomial equation (F5) for the energy. This particular form is also most suited for making a transition to R–S series. Here one makes a particular choice of U, viz.,

$$U = \overline{U} = E - K_0 = \delta E \tag{3.26}$$

$$\epsilon_s(\overline{U}, \overline{U}) \equiv \epsilon_s(\overline{U})$$

$$\overline{U} = \sum_{2}^{2n+1} \epsilon_s(\overline{U})$$

$$= \sum_{2}^{2n+1} \sum_{k=0}^{s-2} \frac{1}{k!} \overline{U}^k \left(\frac{\partial^k \epsilon_s}{\partial \overline{U}^k} \right)_{\overline{U}=0}. \tag{3.27}$$

The energy denominators reduce to the familiar R–S form

$$E - K_m - \overline{U} = K_0 - K_m \equiv (m).$$

One defines some new quantities with a view of decomposing the series (27) into powers of $\overline{U}$,

$$(r, t_1 t_2 \ldots t_{r-1}) = \sum_{m,n,\ldots,y,z\neq 0} \frac{V_{0m} V_{mn} \ldots V_{yz} V_{z0}}{(m)^{t_1} (n)^{t_2} \ldots (z)^{t_{r-1}}} \tag{3.28}$$

V occurs r times in the above expression and the powers t_i take on values 1, 2, 3 . . . .

Then

$$\epsilon_{r,t} = \sum_{t_1,\ldots,t_{r-1}} (r, t_1 t_2 \ldots t_{r-1}) \tag{3.29}$$

where the sum extends over all sets of t_i such that

$$t = \sum_{i=1}^{r-1} t_i > r - 1$$

(e.g., $r = 4$, $t = 6$, the sum includes all permutations of the triplets (4, 1, 1; 3, 2, 1; 2, 2, 2)). We obtain thus a power series in $\overline{U}$ for the general term

$$\epsilon_{s+1}(\overline{U}) = \epsilon_{s+1,s} - \overline{U}\epsilon_{s,s} + \overline{U}^2 \epsilon_{s-1,s} + \ldots + (-)^{s-1} \overline{U}^{(s-1)} \epsilon_{2s}. \tag{3.30}$$

This leads to a polynomial equation in $\overline{U}$ whose lowest root gives the required energy value

$$\overline{U} = \sum_{2}^{2n+1} \epsilon_{s,s-1} - \overline{U} \sum_{2}^{2n} \epsilon_{s,s} + \overline{U}^2 \sum_{2}^{2n-1} \epsilon_{s,s+1} \ldots - \overline{U}^{2n-1} \epsilon_{2,2n}. \tag{3.31}$$

We are still in the framework of the B–W method and we are still considering the energy expression corresponding to the nth order approximant to the *wave function*.

As an aid to actual computation of terms in (31) we obtain a relation between $\epsilon_{s,t}$ for different values of t for a given s. Returning to the original equation for E (before (26)) and (27) we get on putting $U = 0$

$$\overline{U} = \sum_{2}^{2n+1} \epsilon_s(\overline{U}, 0)$$

$$= \exp\left(\overline{U} \frac{\partial}{\partial K_0}\right) \sum_{2}^{2n+1} \epsilon_{s,s-1}. \qquad (3.32)$$

The second line is the Taylor series in powers of $\overline{U}$. The function $\epsilon_{s,s-1}$ arises because it is the same function of K_0 that $\epsilon_s(\overline{U}, 0)$ is of $K_0 + \overline{U}$.

Comparing the coefficients of $\overline{U}$ in (27), (31) and (32)

$$\epsilon_{s,s+k-1} = (-)^k \frac{1}{k!} \frac{\partial^k}{\partial K_0^k} \epsilon_{s,s-1}$$

$$= (-)^k \frac{1}{k!} \frac{\partial^k}{\partial \overline{U}^k} \epsilon_s(\overline{U}) \bigg|_{\overline{U}=0}. \qquad (3.33)$$

Here we have succeeded in expressing the higher terms in t of a given $\epsilon_{s,t}$ as derivatives of the lower t terms for the same s. Thus, for example,

$$\epsilon_{4,6} = -\frac{1}{3} \frac{\partial}{\partial K_0} \epsilon_{4,5} = \frac{1}{3.2} \frac{\partial^2}{\partial K_0^2} \epsilon_{4,4} = -\frac{1}{3!} \frac{\partial^3}{\partial K_0^3} \epsilon_{4,3}$$

$$\epsilon_{4,3} = (4.111)$$

4. RAYLEIGH-SCHRÖDINGER SERIES (F5)

This series can be obtained in various ways. The usual method of associating a parameter with the perturbation V and then expanding the energy and wave function in terms of this parameter is familar from the textbooks. We have given some other methods and have mentioned that the R–S series can be obtained by expanding the energy denominators of the B–W series in successive approximations.

4a. *Relation to B–W series and convergence of the two series*

Here the R–S series is obtained by the formal device of associating an expansion parameter, λ, with W, expressing $\overline{U}$ of the previous section as a power series in λ and finally putting $\lambda = 1$. Letting

$$\overline{U} = \sum \lambda^s \overline{U}_s \tag{4.1}$$

in (3.31) and equating the coefficients we have in successive orders of λ

$$\overline{U}_2 = \epsilon_{21}, \quad \overline{U}_3 = \epsilon_{32}, \quad \overline{U}_4 = \epsilon_{43} - \overline{U}_2\epsilon_{22}$$

$$\overline{U}_5 = \epsilon_{54} - \overline{U}_3\epsilon_{22} - \overline{U}_2\epsilon_{33}, \text{ etc.} \tag{4.2}$$

The finite *approximants* up to a given order p are

$$\overline{U}^{(p)} = \sum_{2}^{2p+1} \overline{U}_s; \; p = \tfrac{1}{2}, 1, \tfrac{3}{2} \ldots \tag{4.3}$$

This approximant is not necessarily the lowest root of the B–W energy equation obtained from the pth order wave functions, for integral p. As a matter of fact much of the connection with B–W wave functions is lost at this point.

It will be shown later that for a large uniform system of A particles interacting through two body forces

$$\lim_{A\to\infty} (\overline{U}_s/A) = f_s(\Omega_0) \tag{4.5}$$

where $f_s(\Omega_0)$ is a constant depending only on the volume Ω_0 per particle and the index s but not on A. It is then clear that terms like $\epsilon_{s,t}$ must be non-linear in A and that in each order of $\overline{U}_s$ an exact cancellation of all powers of A higher than the first should occur. This is the result of the famous 'linked-cluster' expansion which has been arrived at in various ways some of which will be described in succeeding chapters.

Returning to B–W series (3.31), put $n = 1$, then

$$\overline{U} = \frac{\epsilon_{21} + \epsilon_{32}}{1 + \epsilon_{22}}. \tag{4.6}$$

As A becomes very large this tends to a constant independent of A. In order to obtain a $\overline{U}$ proportional to A, such as it should be, one has to take larger and larger values of n as A increases. One feels that $n \simeq A$ should be able to give a good approximation. This makes the B–W method practically useless for large A.

Feenberg has pointed out that one may need a substantially smaller number of terms if use is made of the continued fraction form of this B–W method. No big increase in computational labour is involved. Up to the same order $n = 1$ in the continued fraction form we have from (3.12)

$$\begin{aligned}\bar{U} &= [\epsilon_2(\bar{U})]^2[\epsilon_2(\bar{U}) - \epsilon_3(\bar{U})]^{-1} \\ &= \epsilon_{21}^2[\epsilon_{21} - \epsilon_{32} + \bar{U}\epsilon_{22}]^{-1}. \end{aligned} \tag{4.7}$$

This is quadratic in $\bar{U}$ and has the solution

$$\bar{U} = \frac{\epsilon_{21}}{\epsilon_{22}^{\frac{1}{2}}} \simeq A^{\frac{1}{2}}. \tag{4.8}$$

Hence one may guess that in this method one may get the desired result roughly in the $(A^{\frac{1}{2}})^{\text{th}}$ order.

These considerations are by no means rigorous but they do illustrate the difficulties faced in practical applications of these methods.

4b. *Scale transformation* ($F5$)

It was seen that one could either minimise the energy in the B–W series w.r.t. the parameter of the scale transformation or one could obtain a continued fraction expression for the energy. In either case an improvement over the ordinary B–W series resulted. The same freedom is available for R–S series. However the choice of an optimum value for μ cannot be made through the condition of minimisation as will be clear on inspecting the first few approximants. From equations (2) and (3), on dropping the bars on the U's we have

$$U'^{(\frac{1}{2})} = \frac{1}{\mu}\epsilon_{21} = \frac{1}{\mu}U_2$$

$$U'^{(1)} = \frac{1}{\mu}\left[1 + \frac{\mu - 1}{\mu}\right]U_2 + \frac{1}{\mu^2}U_3$$

$$\begin{aligned}U'^{(\frac{3}{2})} = \frac{1}{\mu}\left[1 + \frac{\mu-1}{\mu} + \left(\frac{\mu-1}{\mu}\right)^2\right]U_2 \\ + \frac{1}{\mu^2}\left[1 + \frac{2\mu - 1}{\mu}\right]U_3 + \frac{1}{\mu^3}U_4.\end{aligned}$$

In general

$$U'_s = \frac{1}{\mu^{s-1}} \sum_{k=0}^{s-2} \binom{s-2}{k} (\mu - 1)^k U_{s-k} \tag{4.9}$$

and, of course,

$$U'^{(p)} = \sum_{2}^{2p+1} U'_s$$

one wants a μ such that even the first few approximants are close to the limit of the sequence. This limit not being known, strictly speaking, the desirable choice of μ cannot be made. But hoping that the analogy with the B–W case may be used one looks for choices of μ that will render the approximants invariant of μ. This is not quite as paradoxical as it seems. For consider a particular choice

$$\mu_i = 1 - U_3/U_2. \tag{4.10}$$

This implies that $U'_3 = 0$. Further

$$U_i^{(\frac{1}{2})} = U_i^{(1)} = \frac{U_2^2}{U_2 - U_3} = U_2 + U_3 + \frac{U_3^2}{U_2 - U_3} \tag{4.11}$$

$$U_i^{(\frac{3}{2})} = \frac{U_2^2}{U_2 - U_3} + \frac{U_2^2(U_2U_4 - U_3^2)}{(U_2 - U_3)^3} \tag{4.12}$$

and

$$U_i^{(\frac{3}{2})} = U_2 + U_3 + U_4 + \text{remainder}. \tag{4.13}$$

In each case the remainder is small if $|U_3/U_2| \ll 1$. Apart from the remainder these are just the original linear approximants to $U^{(1)}$ and $U^{(\frac{3}{2})}$. Now note that

$$\frac{U_2^2}{U_2 - U_3} = \frac{U_2'^2}{U'_2 - U'_3}$$

and

$$\frac{U_2'^2(U'_2U'_4 - U_3'^2)}{(U'_2 - U'_3)^3} = \frac{U_2^2(U_2U_4 - U_3^2)}{(U_2 - U_3)^3} \tag{4.14}$$

which shows that the choice (10) for μ does leave the series invariant.

It is quite clear that here we are on rather dangerous grounds. We would like to say that the invariant forms of these approximants allow a sort of extrapolation between the given ordinary approximants and the series limit. It is not clear whether it can be mathematically proved at all. But one could use them to see how they work in practice and worry about mathematical proofs only if they are otherwise found successful.

This series of papers by Feenberg and his collaborators on this subject are very good examples of the *method of research* advocated by Polya in his very instructive books 'Patterns of Plausible Reasoning' and 'Induction and Analogy in Mathematics'. One should read these papers carefully and closely to see how a wealth of general theorems and ideas arise in mathematics by working through many particular examples, always, of course, with an eye on the form and a feeling for the generalisable. I have collected here only some of the results of these investigations. There is a great deal of material in these articles which will also benefit those interested in actual computational work.

To end this discussion of the R–S method we just mention a work by Speisman (S2) who has given a convergence condition for the R–S series expressed as an upper limit on the strength of the interaction operator. As Feenberg remarks one could perhaps relate the scale transformation to this condition to provide a suitable choice for μ.

5. SUM RULE TECHNIQUES

An innovation suggested originally by Dalgarno and Lewis (D1, D2) has been somewhat extended and used by Schwartz (S3, S4) in computations of two electron problems and the lamb shift in hydrogen. The method makes use of the (differential) Schrödinger equation itself to express higher order correction terms of the perturbation series by single integrals and is claimed to be very powerful. At the moment it is not clear whether it can be adapted for handling many particle systems. However we give here a summary of their work.

The important step is to note the identity

$$\frac{\langle p_a|[K_{\text{op}}, f]|p_0\rangle}{K_a - K_0} = \langle p_a|f|p_0\rangle \tag{5.1}$$

where

$$[A, B] = AB - BA$$

for any operator f which does not commute with K_{op}. This enables one to write

$$\frac{\langle p_a|V|p_0\rangle}{K_a - K_0} = \langle p_a|f|p_0\rangle \tag{5.2}$$

provided an operator f could be found such that

$$V = [K_{op}, f]. \tag{5.3}$$

Using the relation (3) the second order correction to the energy is given by

$$\begin{aligned} E^{(2)} &= \sum_n \frac{\langle p_0|V|p_n\rangle \langle p_n|[f, K_{op}]|p_0\rangle}{K_n - K_0} \\ &= \sum_{n\neq 0} \langle p_0|V|p_n\rangle \langle p_n|f|p_0\rangle \\ &= \langle p_0|Vf|p_0\rangle - E^{(1)}\langle p_0|f|p_0\rangle \end{aligned} \tag{5.4}$$

and the first order correction to the wave function by

$$\begin{aligned} |F_1\rangle &= \sum_{n\neq 0} |p_n\rangle \frac{\langle p_n|V|p_0\rangle}{K_0 - K_n} \\ &= f|p_0\rangle - \langle p_0|f|p_0\rangle \cdot |p_0\rangle. \end{aligned} \tag{5.5}$$

The operator equation (3) is rather difficult to solve in practice. The above relations (4) and (5) are satisfied even if f obeys the weaker condition

$$[f, K_{op}]|p_0\rangle = (V - E^{(1)})|p_0\rangle. \tag{5.6}$$

Some of the methods for solving this equation have been pointed out by Schwartz (S3, S4). These are:

1. When f is a function of space coordinates only and

$$K_{op} = -\frac{\hbar^2}{2m}\nabla^2 + v(\mathbf{r}) \tag{5.7}$$

then one has to solve the inhomogeneous differential equation

$$\frac{\hbar^2}{2m}[\nabla^2 f + 2\nabla f\cdot\nabla]\Phi_0 = (V - E^{(1)})\Phi_0 \tag{5.8}$$

where $\Phi_0 = |p_0\rangle$. The boundary conditions on f are provided by the corresponding conditions on the perturbed wave function $|\Psi\rangle \simeq |\phi_0\rangle + |F_1\rangle$. For the one-dimensional case the above equation is solvable by quadratures.

2. A variational principle may be constructed for the second order energy and for f. The quantity

$$J \equiv \langle\phi_0|\{\bar{f}(V - E^{(1)}) + (V - E^{(1)})\bar{f} - \tfrac{1}{2}[\bar{f}, [\bar{f}, H_0]]\}|\phi_0\rangle \tag{5.9}$$

is stationary w.r.t. arbitrary variations of the operator $\bar{f}$ provided $\bar{f}$ satisfies an equation like (6) and its conjugate. At its extremum $J = E^{(2)}$. Or put $\bar{f} = cf'$ and vary w.r.t. c where c is a constant. Then at the minimum

$$\begin{aligned} c &= I_1/I_2;\ J = \tfrac{1}{2}[I_1^2/I_2] \\ I_1 &= \langle\phi_0|f'(V - E^{(1)}) + (V - E^{(1)})f'|\phi_0\rangle \\ I_2 &= \langle\phi_0|[f', [f', K_{op}]]|\phi_0\rangle. \end{aligned} \tag{5.10}$$

The energy is not very sensitive to the precise form of f'; often an approximate form of f suffices for this purpose.

Extension to higher orders (D1, D2) may be made by repeatedly using the basic identity (1). Alternatively one may use a method due to Dalgarno and Stewart (D2): Associating the usual parameter λ with the perturbing potential V we can write the perturbed wave functions and the energy

$$|\Psi\rangle = \sum_{n=0}^{\infty} \lambda^n |F_n\rangle,\ E = \sum_{n=0}^{\infty} \lambda^n E^{(n)}. \tag{5.11}$$

Substituting in the Schrödinger equation

$$(H - E)|\Psi\rangle = (K_{op} + \lambda V - E)|\Psi\rangle = 0 \tag{5.12}$$

and equating the coefficients of the various powers of λ separately to zero we have

$$(K_{op} - K_0)|F_0\rangle = 0;\ |F_0\rangle \equiv |\phi_0\rangle \tag{5.13}$$

$$(K_{op} - K_0)|F_1\rangle + (V - E^{(1)})|F_0\rangle = 0 \tag{5.14}$$

which is equivalent to the equations (5) and (6). For the nth order, where $n > 1$

$$(K_{op} - K_0)|F_n\rangle + (V - E^{(1)})|F_{n-1}\rangle = \sum_{r=2}^{n} E^{(r)}|F_{n-r}\rangle. \tag{5.15}$$

(The usual perturbation formulae are obtained by writing $|F_n\rangle = \sum_r a_{nr}|p\rangle$ etc..)

These are recursion relations between quantities of different orders. The nth order energy is

$$E^{(n)} = \langle F_0|(V - E^{(1)})|F_{n-1}\rangle - \sum_{r=1}^{n-2} E^{(r+1)}\langle F_0|F_{n-r-1}\rangle. \quad (5.16)$$

We can now use Eq. (15) to reduce the orders of the F's. The process ends when the highest order of F entering the equation is $\frac{1}{2}n$ for even n and $(n - 1)/2$ for odd n. This leads to the remarkable theorem (Dalgarno and Stewart (D2)) that the *energy can be evaluated to* $(2s + 1)^{\text{th}}$ *order provided that the wave function is known to the* s^{th} *order.*

Matrix elements of an arbitrary operator Q: We illustrate further by applying the sum rule technique to the diagonal element w.r.t. first order wave function

$$|\Psi^{(1)}\rangle = |p_0\rangle + |F_1\rangle$$

$$\langle\Psi^{(1)}|Q|\Psi^{(1)}\rangle = \langle p_0|Q|p_0\rangle + 2\sum_{s\neq 0} \frac{\langle p_0|V|p_s\rangle\langle p_s|Q|p_0\rangle}{K_0 - K_s}. \quad (5.17)$$

The identity (1) may be used to replace either of the matrix elements. We have

$$\langle\Psi^{(1)}|Q|\Psi^{(1)}\rangle = (1 - 2\langle p_0|f|p_0\rangle)\langle p_0|Q|p_0\rangle + 2\langle p_0|fQ|p_0\rangle \quad (5.18)$$

where f satisfies (6). Or we may have

$$\langle\Psi^{(1)}|Q|\Psi^{(1)}\rangle = (1 - 2\langle p_0|g|p_0\rangle)\langle p_0|Q|p_0\rangle + 2\langle p_0|Vg|p_0\rangle \quad (5.19)$$

where g satisfies the equation

$$[g, K_{\text{op}}]|p_0\rangle = (Q - Q^{(1)})|p_0\rangle;\ Q^{(1)} = \langle p_0|Q|p_0\rangle. \quad (5.20)$$

If these equations for f and g are not exactly solvable, a possible confidence test on the approximate evaluation would be to compare the two values (18) and (19).

Further, some relations between parts of the matrix elements of an operator can be established. In particular, it may be proved that for an atom in the fixed nucleus approximation the total potential and the kinetic energies can be calculated to $(2n + 1)^{\text{th}}$ order if the wave function is known only to n^{th} order (D1, D2).

6. COUNTING OPERATOR METHODS

Recall the integral equation for M derived previously

$$M = 1 + \frac{1}{E - K_{\text{op}} - O}[V - O]M. \tag{6.1}$$

One can carry out various formal manipulations on this equation to extract a solution. One way of doing this is to introduce (R1) a new operator, P, such that

$$M = 1 + \frac{1}{E - K_{\text{op}} - O} PVM. \tag{6.2}$$

It follows that the operators P and O should satisfy the equation

$$(1 - P)VM|p_0\rangle = OM|p_0\rangle. \tag{6.3}$$

This condition is weaker than the *operator relation* $(1 - P)VM = OM$ which would also imply the equivalence of (1) and (2). The procedure for solving the integral equation will now be to select an O such that the propagator $E - K - O$ has the right order of magnitude and then choose P to put the series expansion in a useful form. As a simple example note that $P = 1 - \Lambda_{p_0}$ gives the B–W series. The method assumes new significance with the use of the so-called 'counting operators' for P. In one of its forms it leads to the formulae encountered in the multiple scattering formulation of the optical model (F13, W2, W3). That apparently is the reason why Riesenfeld and Watson (R1) call these methods 'optical model' methods. We consider two important specific cases.

It will be seen that these counting operators are of a highly unusual nature. In particular they seem to be non-associative and their use gives rise to certain ambiguities.

6a. *Feenberg's method*

Define $P = P_{\text{F}}$, the Feenberg operator, to be such that in the expansion of $\langle p'|P_{\text{F}}VM|p\rangle$ the state p' is not equal to any subsequent intermediate state. The meaning of this is clarified by the following evaluation of the 'Feenberg potential' V_{F}, a diagonal matrix one of whose elements is the level shift. Let

$$P = P_{\mathrm{F}}$$

$$a = E - K - O \tag{6.4}$$

$$O = V_{\mathrm{F}}. \tag{6.5}$$

Since O is a diagonal operator

$$\langle p_n|O|p_m\rangle = \langle p_n|VM|p_n\rangle\delta_{mn}. \tag{6.6}$$

We have to show that these definitions of O and P are consistent, that is they satisfy equation (3). Consider the diagonal element

$$\begin{aligned} V_{\mathrm{F}}(p_0) &= \langle p_0|VM|p_0\rangle \\ &= \langle p_0|\left\{V + V\frac{1}{a}P_{\mathrm{F}}V + V\frac{1}{a}P_{\mathrm{F}}V\frac{1}{a}P_{\mathrm{F}}V + \ldots\right\}|p_0\rangle \\ &= V_{00} + \sum_n V_{0n}\left(\frac{1}{a}P_{\mathrm{F}}V\right)_{n0} + \sum_{n,m} V_{0n}\left(\frac{1}{a}P_{\mathrm{F}}V\right)_{nm}\left(\frac{1}{a}P_{\mathrm{F}}V\right)_{m0} + \ldots \\ &= V_{00} + \sum_{n\neq 0}\frac{V_{0n}V_{n0}}{E - K_n - O_{nn}} \\ &\quad + \sum_{\substack{m,n\neq 0\\ n\neq m}}\frac{V_{0n}V_{nm}V_{m0}}{(E - K_n - O_{nn})(E - K_m - O_{mm})}. \end{aligned} \tag{6.7}$$

Note that in the expansion of O_{nn} again certain intermediate states do not occur

$$\begin{aligned} O_{nn} = V_{nn} &+ \sum_{m\neq n}\frac{V_{nm}V_{mn}}{(E - K_m - O_{mm})} \\ &+ \sum_{\substack{m,p\neq n\\ m\neq p}}\frac{V_{nm}V_{mp}V_{pn}}{(E - K_m - O_{mm})(E - K_p - O_{pp})} + \ldots \end{aligned} \tag{6.8}$$

It is desirable to introduce a notation which explicitly exhibits the states omitted from the energy denominators. We write (F6, M1)

$$(k^2)_{pqr\ldots s;n} = \{K_n + O_{nn} \text{ with } n \neq p, q, r \ldots, s\} \tag{6.9}$$

then

$$\begin{aligned} E - K_0 = \Delta E &= \langle p_0|VM|p_0\rangle \\ &= V_{00} + \sum_{0\neq n}\frac{V_{0n}V_{n0}}{E - (k^2)_{0;n}} \\ &\quad + \sum_{\substack{m,n\neq 0\\ m\neq n}}\frac{V_{0n}V_{nm}V_{m0}}{(E - (k^2)_{0;n})(E - (k^2)_{0n;m})} + \ldots. \end{aligned} \tag{6.10}$$

with any arbitrary state p_1.

$$\langle p_1|P_F VM|p_0\rangle = \langle p_1|\left\{P_F V + P_F V\frac{1}{a}P_F V + \ldots\right\}|p_0\rangle$$

$$= V_{10} + \sum_{n\neq 0,1} \frac{V_{1n}V_{n0}}{E-(k^2)_{1;n}}$$

$$+ \sum_{\substack{m,n\neq 0,1\\ m\neq n}} \frac{V_{1n}V_{nm}V_{m0}}{(E-(k^2)_{1;n})(E-(k^2)_{1n;m})} + \ldots$$

Or, with injunctions on indices inderstood and $(k^2)_1 = K_1 + O_{11}$

$$\langle p_1|(1-P_F)VM|p_0\rangle = \frac{V_{11}V_{10}}{E-(k^2)_1} + \sum \frac{V_{11}V_{1m}V_{m0}}{(E-(k^2)_1)(E-(k^2)_{1;m})}$$

$$+ \sum \frac{V_{1n}V_{n1}V_{10}}{(E-(k^2)_{1;n})(E-(k^2)_1)} + \ldots$$

$$= \langle p_1|V|p_1\rangle \langle p_1|\left\{1 + \frac{1}{a}P_F V + \frac{1}{a}P_F V\frac{1}{a}P_F V + \ldots\right\}|p_0\rangle$$

$$+ \langle p_1|V\frac{1}{a}P_F V|p_1\rangle\langle p_1|\left\{1 + \frac{1}{a}P_F V + \frac{1}{a}P_F V\frac{1}{a}P_F V + \ldots\right\}|p_0\rangle + \ldots$$

$$= \langle p_1|VM|p_1\rangle \langle p_1|M|p_0\rangle$$

or

$$\langle p_1|(1-P_F)VM|p_0\rangle = V_F(p_1)\langle p_1|M|p_0\rangle. \qquad (6.11)$$

Or recalling the definition of O

$$\langle p_1|(1-P_F)VM|p_0\rangle = \langle p_1|OM|p_0\rangle$$

which shows that equation (3) is satisfied. We have shown that the definitions of P_F and O, equation (4) to (7), are consistent and they yield the (implicit) Feenberg formula (10) for the perturbed energy.

For a discussion of the convergence properties of this series see (F6, M1). We might add, without proof, that the series still has a finite radius of convergence in terms of the strenth of the perturbation V.

6b. *Multiple scattering solutions*

We are discussing various forms of the perturbation theory in order to indicate what sort of methods have been used to put the expansions in forms most suitable for actual computations. In the present section a very concise derivation of the multiple scattering approach to energy eigenvalue problem and the optical model potential is given (F13, W2, W3 and their various sequels; more recent applications are considered in reference R5). The brevity of this section should not obscure its great importance. We shall later discuss in Ch. II and III, alternative derivations of the 'linked cluster' and t-matrix formulae and then their meaning will become more transparent.

Let α be the index for a pair, then restricting ourselves to a pure two-body interaction, we have

$$V = \sum_{\alpha} v_{\alpha}. \tag{6.12}$$

Although we do not wish to prejudice the reader at this stage by appealing to physical intuition, we may remark that now we want to separate, in a rather loose sense, the interaction effects of one pair of particles from those due to the other pairs. In the last term of (2) we have $\sum v_{\alpha}M$ where each term $v_{\alpha}M$ is effected by all the pairs. We write

$$v_{\alpha}M = t_{\alpha}M_{\alpha} \tag{6.13}$$

where we want to define the quantities on the right hand side in such a way that t_{α} represents the major effects of the pair α and M_{α} those due to the rest. We have

$$M = 1 + \frac{1}{a}\sum_{\alpha} Pt_{\alpha}M_{\alpha} \tag{6.14}$$

and observe that in the special case when only one of the v_{α}'s is non-vanishing we can choose $M_{\alpha} = 1$ so that the solution for this case is provided by $M = 1 + (1/a)Pt_{\alpha}$ which from (6.13) leads to a natural definition of t_{α}, the so-called two-body reaction matrix,

$$t_{\alpha} = v_{\alpha}\left(1 + \frac{1}{a}Pt_{\alpha}\right). \tag{6.15}$$

This definition may be carried over to the general case where, on using (6.15) in (6.13), we have the general expression for M_{α}

$$M_{\alpha} = M - \frac{1}{a}Pt_{\alpha}M_{\alpha} = 1 + \sum_{\beta\neq\alpha}\frac{1}{a}Pt_{\beta}M_{\beta}. \tag{6.16}$$

All reference to v_α in the level shift expression (2.12) may be eliminated. It can be replaced everywhere by the *pseudo two-body* reaction matrix, t_α. The equations (14), (15) and (16) thus provide a rigorous (formal) solution of the problem with the level shift given by

$$V_c(p_0) = \langle p_0 | \sum_\alpha t_\alpha M_\alpha | p_0 \rangle D(p_0). \tag{6.17}$$

To get actual results from this we have to provide a consistent and useful definition of the operators P and O which are so far unspecified.

The choice of P which leads to the optical model potential expressions of reference (F13, W2) and the energy eigenvalue problem of Brueckner (B11, B12) is

$$P = [1 - \Lambda_{p_0}] P_{\mathrm{ND}} \tag{6.18}$$

where P_{ND}, a matrix with vanishing diagonal elements, is defined by the equations

$$\langle p_a | P_{\mathrm{ND}} t_\alpha | p_b \rangle = (1 - \delta_{ab}) \langle p_a | t_\alpha | p_b \rangle. \tag{6.19}$$

In previous work (B11, B12, F13, W2) the notation

$$P_{\mathrm{ND}} t_\alpha = I_\alpha \tag{6.20}$$

has been used. Neglecting effects of order of $1/N$ we have

$$D(p_0) = 1 \tag{6.21}$$

$$V_c = \langle p_0 | \sum_\alpha t_\alpha M_{0\alpha} | p_0 \rangle \tag{6.22}$$

$$M_{0\alpha} = 1 + \frac{1}{e} \sum_{\beta \neq \alpha} [1 - \Lambda_{p0}] I_\beta M_{0\beta} \tag{6.23}$$

$$e = E - K - \sum_\alpha \langle p | t_\alpha | p \rangle, \tag{6.24}$$

where the subscript 0 on $M_{0\alpha}$ is to remind ourselves that this operator has been constructed specifically for the states $|p_0\rangle$.

In the first approximation of this procedure the energy is given by

$$E_{\lambda 0} = K(p_0) + \sum_\alpha \langle p_0 | t_\alpha | p_0 \rangle. \tag{6.25}$$

This is the Brueckner expression (B11) for the energy in terms of the reaction matrix. In its evaluation the difficulty of non-linear N-

dependence [11] of some terms again arises because of the propagator as discussed in earlier sections.

The anomaly is removed by redefining P in a suitable way quite analogous to Feenberg's operator discussed above. This process is referred to as 'summation over repeated linked clusters' in the literature (Ch. II) and was first carried out by Brueckner (B11) in a different manner. We put $P = P_0$ where P_0 eliminates the diagonal elements of the clusters of t_α's. The effect of this operator is exhibited by the following equation (cf., the derivation of Eq. (11)):

$$\langle\phi_1|(1 - P_0)VM|\phi_0\rangle$$

$$= \langle\phi_1|(1-P_0)\left\{\sum_\alpha t_\alpha\left[1+\sum_{\alpha_1\neq\alpha}\frac{1}{a}P_0t_{\alpha1}+\sum_{\alpha_2\neq\alpha_1\neq\alpha}\frac{1}{a}P_0t_{\alpha1}\frac{1}{a}P_0t_{\alpha2}+\ldots\right]\right\}|\phi_0\rangle$$

$$= \langle\phi_1|\sum_\alpha t_\alpha|\phi_1\rangle\langle\phi_1|\left\{1+\sum_{\alpha_1\neq\alpha}\frac{1}{a}P_0t_{\alpha1}+\ldots\right\}|\phi_0\rangle$$

$$+\langle\phi_1|\sum_\alpha t_\alpha\frac{1}{a}P_0\sum_{\alpha_1\neq\alpha}t_{\alpha_1}|\phi_1\rangle\langle\phi_1|\left\{1+\sum_{\alpha_2\neq\alpha_1}\frac{1}{a}P_0t_{\alpha1}+\ldots\right\}|\phi_0\rangle+\ldots$$

$$= \langle\phi_1|\sum_\alpha t_\alpha|\phi_1\rangle\langle\phi_1|M_\alpha|\phi_0\rangle$$

$$+\langle\phi_1|\sum_\alpha t_\alpha\frac{1}{a}P_0\sum_{\alpha_1\neq\alpha}t_{\alpha_1}|\phi_1\rangle\langle\phi_1|M_{\alpha_1}|\phi_0\rangle+\ldots \tag{6.26}$$

Strictly speaking it is not possible to reduce this series any further. One now makes the observation that M_α differs from M, Eq. (14), only in summation over one pair hence for a system with a large number of pairs it should be possible, in a certain approximation, to write M for each M_α. Hence

$$\langle\phi_1|(1 - P_0)VM|\phi_0\rangle = \langle\phi_1|\left\{\sum_\alpha t_\alpha+\sum_{\alpha\neq\alpha_1}t_\alpha\frac{1}{a}P_0t_{\alpha1}+\ldots\right.$$

$$\left.+\ldots\right\}|\phi_1\rangle\langle\phi_1|M|\phi_0\rangle = \langle\phi_1|V_0M|\phi_0\rangle \tag{6.27}$$

[11] For optical model problem where one considers the interaction of one particle with a medium containing N particles the number of pairs to be summed over is only N and *not* N^2 so that in this case there is no unlinked cluster problem.

where V_0 is a diagonal operator defined by the terms in the curly brackets. We choose $O = V_0$ and thus satisfy equation (3). Finally the infinite expansion of (22) and (23) gives

$$V_0|p_0\rangle = V_c|p_0\rangle \tag{6.28}$$

hence

$$D = 1 \text{ and } M = \boldsymbol{M}. \tag{6.29}$$

6c. *Addition theorem for the energy of a composite system*

As the last topic in the formal perturbation theory we give below the addition theorems of Riesenfeld and Watson (R1) which concern the relationship between energies of parts of a given system.

Here we consider a general problem which reduces in one particular instance to the problem of finding the energy of a single particle moving in a medium such that the total energy may be expressed as a sum of such single particle energies.

Let there be two ways of dividing up the Hamiltonian H. We shall drop the operator subscript and write $K_{op} \equiv K$.

$$\begin{aligned} H &= K + V \\ &= K' + V' \end{aligned} \tag{6.30}$$

with $V = V_1 + V_2$ with $V_1 = V'$ and

$$K' = K + V_2.$$

Assume that the complete set of eigenfunctions $|p\rangle$ and $|p'\rangle$ belonging respectively to K and K' is known and that a one-to-one correspondence between the two sets of eigenfunctions has been established. The eigenvalue problem of H can be formulated in terms of either of these sets; and we wish to investigate the connection between the two.

The following symbols will be needed:

$$\delta E = E_{\lambda 0} - K(p_0) = \lim_{E \to E_{\lambda 0}} \langle p_0 | V\boldsymbol{M} | p_0 \rangle$$

$$\delta E_1 = E_{\lambda 0} - K'(p'_0) = \lim_{E \to E_{\lambda 0}} \langle p'_0 | V\boldsymbol{M}_1 | p'_0 \rangle$$

$$\delta E_2 = \delta E - \delta E_1$$

$a_0, a_1, a_2 \ldots$ arbitrary constants

$\Delta_0 = \delta E - (E - E_{\lambda 0})a_0$

$\Delta_i = \delta E_i - (E - E_{\lambda 0})a_i;\ i = 1$ or 2

$\Delta_2 = \Delta_0 - \Delta_1;\ a_2 = a_0 - a_1$

$a = E - K - \Delta_0$

$d = E - K' - \Delta_1$

W_{d_1} – the Green function in $|p'\rangle$ representation

$\Psi_{\lambda 0}(p_0),\ \Psi_{\lambda 0}(p_0')$ – the eigenfunctions of H in p- and p'-representation respectively.

$$m_2 = 1 + \frac{1}{a}(V_2 - \Delta_2)m_2 = 1 + \frac{V_2 - \Delta_2}{d}$$

$$= \frac{1}{d}a.$$

It follows from the work in section 2, refer especially to Eqs. (2.19 to 2.24), that the solution in the two representations may be expressed by means of the corresponding M matrices.

In the *p-representation*

$$M = 1 + \frac{1}{a}[V_1 + V_2 - \Delta_0]M$$

$$\boldsymbol{M} = MD$$

$$D = \lim_{E \to E_{\lambda 0}} [E - K - \Delta_0]^{-1} W_d^{-1} = |\Psi_{\lambda 0}(p_0)|^{-2}\,(1 - a_0)^{-1}$$

or alternatively in the *p′-representation*

$$M_1 = 1 + \frac{1}{d}[V_1 - \Delta_1]M_1$$

$$\boldsymbol{M}_1 = M_1 D_1$$

$$D_1 = \lim_{E \to E_{\lambda 0}} [E - K' - \Delta_1]^{-1} W_{d_1}^{-1} = |\Psi_{\lambda 0}(p_0')|^{-2}\,(1 - a_1)^{-1}.$$

The limiting process is necessary since the last step in these equations is valid only in the limit. The connection between the two M's is established as follows: Rewritting $1/d$ as $m_2\ (1/a)$ in expression for M_1

$$M_1 = 1 + m_2(1/a)(V_1 - \Delta_1)M_1$$

$$M_1 m_2 = m_2(1 + 1/a(V_1 - \Delta_1)M_1 m_2) \equiv M.$$

The last identity follows by using those integral equations for m_2 and M_1 which involve $1/a$ as the propagator. Since $dm_2 = a$ we have *with all limits taken as* $E \to E_{\lambda 0}$

$$[E - K - V_2 - \Delta_1]m_2|p_0\rangle = \lim a|p_0\rangle = 0.$$

Hence

$$\lim [K + V_2]m_2|p_0\rangle = \lim [E - \Delta_1]m_2|p_0\rangle$$
$$= (E_{\lambda 0} - \delta E_1)|p_0'\rangle.$$

The last step follows because

$$\lim m_2 = \lim\left[1 + \frac{1}{E - K - \Delta_0}[V_2 - \Delta_2]m_2\right]$$
$$= 1 + \frac{1}{K'(p_0) - K - \Delta_2}[V_2 - \Delta_2]m_2;\ \lim \Delta_1 = \delta E_1$$

which, from (2.24), is the integral equation which establishes the correspondence between the p- and p'-representation i.e., solves the eigenvalue problem for K' in terms of the solutions of K (compare also (2.29)).

If the limit exists then

$$\lim m_2|p_0\rangle = |p_0'\rangle C$$

where the normalization constant C is determined as follows:

$$C = \lim \langle p_0'|m_2|p_0\rangle = \lim \langle p_0'|(1/d)a|p_0\rangle$$
$$= \langle p_0'|p_0\rangle \lim \{[E - K(p_0) - \delta E_2 - \Delta_1]^{-1}[E - K(p_0) - \Delta_0]\}$$

and using the definitions of Δ_0 and Δ_1

$$C = \left[\frac{1 - a_0}{1 - a_1}\right]\langle p_0'|p_0\rangle.$$

Finally we calculate the quantity [12]

$$\langle p_0|V_1\boldsymbol{M}|p_0\rangle = \langle p_0|V_1M_1m_2|p_0\rangle D$$
$$= \langle p_0|V_1\boldsymbol{M}_1|p_0'\rangle(D/D_1)C$$
$$= \langle p_0|V_1\boldsymbol{M}_1|p_0'\rangle\langle p_0'|p_0\rangle\left|\frac{\Psi_{\lambda 0}(p_0')}{\Psi_{\lambda 0}(p_0)}\right|^2. \qquad (6.31)$$

This is a rather complex relation showing a connection between the two ways of calculating the effects of the same operator (cf., expressions for δE_1 and δE). It is to be noted that any reference to the arbitrary constants a_0 and a_1 has been eliminated.

[12] In the following relation one can replace $\langle p_0|V_1$ by any arbitrary operator.

6c (i). CONNECTION WITH THE OPTICAL MODEL PROBLEM

The addition theorem described above has immediate application to the problem of finding an expression such that the energy of an N-particle system can be expressed as a 'sum' of an $(N-1)$-particle and a one-particle system. One defines

$$V_1 \to V_i \equiv \sum_{s\neq i}^{N} V_{is}$$

$$V_2 \to V - V_i$$

$$K' \to K + V_2 \equiv K + V - V_i.$$

The eigenfunctions of K' are denoted by $|p'(i)\rangle$.

The extra energy of adding one more particle is

$$\delta E_i = \lim \langle p_0'(i)\,|V_i \boldsymbol{M}(i)|\,p_0'(i)\rangle.$$

The B–W expression for $M(i)$ is (see (2.26))

$$\boldsymbol{M}(i) = 1 + [E - K - (V - V_i)]^{-1}[1 - \Lambda_i]V_i\boldsymbol{M}(i)$$

where Λ_i is the projection operator on to the state $|p_0'(i)\rangle$. One can, on the other hand, write down the multiple scattering solution as before (see (16), (23) and (24))

$$V_i\boldsymbol{M}(i) = \sum_{s\neq 1}^{N} t_{is}M_{si} \tag{6.32a}$$

$$M_{si} = 1 + \frac{1}{e}[1 - \Lambda_i] \sum_{q\neq s,i} I_{iq}M_{qi} \tag{6.32b}$$

where

$$e = E - K' - \sum_{s\neq i}^{N} \langle p_0'(i)\,|t_{is}|\,p_0'(i)\rangle. \tag{6.32c}$$

The t and I matrices are as defined before but the projection and P_{ND} operators now act on $p_0'(i)$-states. On comparing with the multiple scattering solution of the energy eigenvalue problem Eq. (21) to (25), one sees the formal equivalence of the two problems. The difference occurs only in the boundary conditions, the optical model usually being the potential seen by an unbound particle passing through the medium.

6c (ii). THE TOTAL ENERGY OF AN N-BODY SYSTEM

$$E_{\lambda 0} = K(p_0) + \delta E$$

$$\delta E = \lim \langle p_0 | V M | p_0 \rangle$$

$$= \tfrac{1}{2} \sum_{i=1}^{N} \{\lim \langle p_0 | V_i M | p_0 \rangle\}.$$

Using (31) with appropriate change of notation

$$\delta E = \tfrac{1}{2} \sum_{i=1}^{N} \lim \left\{ \langle p_0 | V_i M(i) | p_0'(i) \rangle \langle p_0'(i) | p_0 \rangle \left| \frac{\Psi_{\lambda 0}(p_0'(i))}{\Psi_{\lambda 0}(p_0)} \right|^2 \right\}. \tag{6.33}$$

To evaluate this the expression for $V_i M(i)$ given in (32) is used. The wave functions $\Psi_{\lambda 0}$ and $|p_0'(i)\rangle$ are to be evaluated from the appropriate $\boldsymbol{M}$ matrices in a self-consistent manner. The propagator has no anomalous N-dependence as long as one does not mix the terms of various expansions, i.e., if each of the quantities is calculated separately to a given order and then put into the above equation. This is one way of removing the difficulty associated with anomalous N-dependence of the many body perturbation series. In practice the most successful method of achieving this end has been to perform summations in each order of the R–S series to eliminate linked clusters. This will be taken up in succeeding chapters. The equivalence of the present method to that of Brueckner can be shown for the first few orders, but for higher orders it becomes complicated.

LIST OF IMPORTANT SYMBOLS IN CHAPTER I

A — number of particles in the system. Used interchangeably with N

E — a complex number which in special cases $E = E_\lambda$ becomes the perturbed energy. Denoted also by z or ζ in Chapter II

E_λ — energy of the total system, the perturbed energy. Denoted by $\mathscr{E}_\lambda$ or just $\mathscr{E}$ in other chapters

$E^{(r)}$ — r^{th} order Rayleigh-Schrödinger energy correction, (5.11)

$|F_r\rangle$ — r^{th} order Rayleigh-Schrödinger wave function correction, (5.11)

G_i — variational parameters

H — Hamiltonian of the total system, the perturbed Hamiltonian

I_α — incoherent part of the t-matrix, (6.20)

K_l — eigenvalue of unperturbed Hamiltonian. Denoted by E_a or E_α in other chapters

K_{op} — unperturbed Hamiltonian. Denoted by H_0 in other chapters

M — wave-matrix (2.20) after factoring out the diagonal part

$M^{(n)}$ — n^{th} approximation to M-matrix

M — the formally exact wave-matrix, (2.6)

$M_k(i)$ — i^{th} order of part of a matrix element of M in p-representation, (3.6′)

N — number of particles in the system. Used interchangeably with A

O — an arbitrary operator

$|p_n\rangle$ — unperturbed eigenvector (of K_{op}). Also denoted by Φ_n

P — an arbitrary operator

$P_{\text{F}}, P_{\text{ND}}$ — counting operators (section 6)

t_α — the reaction operator corresponding to the αth pair. The matrix of this operator is the t-matrix

U — the parameter of the translation of origin transformation, (3.26)

$\overline{U}$ — a particular value of U, (3.26)

$\overline{U}_s$ — s^{th} order part of $\overline{U}$, (4.1)

$\overline{U}^{(p)}, (\equiv U^{(p)})$ — The pth approximation of $\overline{U}$, (4.3)

$U'^{(p)}$, U'_s	quantities belonging to arbitrary scale μ, (4.9)
$U_i^{(p)}$	quantities belonging to particular scale μ_i, (4.11) etc.
V	the perturbation (potential)
V_{lk}	matrix element of V between unperturbed states l and k
$W(E)$	Green's function of the total system. The negative of this quantity denoted by $R(z = E)$ is called the resolvent operator in Chapter II
W_d	diagonal part of the Green's function in the p-representation
W	perturbing potential after the transformation
Δ	arbitrary operator
δE	level shift corresponding to the state p_0 ($\equiv V_c(p_0)$)
ϵ_i	i^{th} energy term in the B–W expansion, (3.3) and (3.5)
$\epsilon_{r,t}$	terms in the power series for ϵ_r, (3.30)
Λ_p	the projection operator onto the state $\|p\rangle$
μ	parameter of scale transformation
Φ_l	unperturbed eigenfunctions (of K_{op}), also denoted by $\|p_l\rangle$
Ψ_λ	perturbed eigenfunctions (of H)
$\Psi_\lambda(p)$ ($\equiv \langle p\|\Psi_\lambda\rangle$)	amplitude of the state p in Ψ_λ
$\Psi_\lambda^{(n)}$	n^{th} approximant to Ψ_λ

See the list on p. 37 and 38 for symbols used in section 6c.

CHAPTER II

DIAGRAMMATIC METHODS, LINKED CLUSTER [13] THEOREM AND GENERAL FORMULAE

1. INTRODUCTION

In the previous chapter we discussed various perturbation methods by purely algebraic means, in order, among other things, to show the connection between different methods and to emphasise the importance of the questions of convergence. The statistics obeyed by the particles and for the most part, the nature of the interaction between them played no role in that discussion. In studying an actual system one has to specify these properties. As a result there arise questions concerning the *structure of the terms*. Diagrammatic methods have been devised mainly for dealing with such questions. The nature of a diagram, of course, depends on the type of interaction. In this book, since we are concerned only with the nuclear problem, we shall consider only static two-body forces [14] between fermions. The formal results obtained here will, therefore, also be applicable to other such systems like electrons in metals, liquid He^3 and liquid N^{15}.

The most satisfactory way of introducing the diagrams is through the second quantisation methods of sec. 3 and 4 of this chapter. However, in order to effect a smooth transition from the methods of the previous chapter, we study in the next section the structure of the first few terms of the R–S series using unperturbed determinantal wave functions appropriate to the Fermi-Dirac statistics. Diagram-

[13] The term 'clusters' has also been used in other connections, for instance, there are the cluster expansions of Mayer (M10) in statistical mechanics, the cluster expansions of Iwamoto and Yamada (I1, T4) for energies of quantum mechanical systems, and the particle cluster model of Wildermuth and Kanellopoulos (W5). One should guard against confusion; in particular, there is no obvious connection between the occurrence of the present cluster terms and the possible existence of alpha-particles or other such clusters in nuclei.

[14] Diagrams for non-static interaction and mixed Fermion-Boson systems are well known in field theory and may be studied from any of the several text books now available .Three-body interactions have been discussed by Van Hove (v.H2) in connection with a solid state problem. Polkinghorne (P2) has shown that by suitable redefinition some of the effects of three-body interactions may be absorbed in the expressions obtained for two-body interactions.

matic representation of the matrix elements will be introduced here as a mnemonic device for keeping track of the complex structure of the terms. In the third order, for the first time, there arise certain cancellations which are explicitly shown and the linked cluster theorem is stated. This treatment is designed to emphasise the fact that all the subsequent developments are essentially based on the R–S series. Also it seemed easier to introduce diagrams in this way and it is hoped that it will reduce somewhat the awe in which the diagrams are held and the glibness with which they are often used. These points are not easy to make in the second quantised treatment. As a by-product we have given a complete enumeration of the terms of the third order and exhibited the form of the expressions which will have to be evaluated if present calculations of the many body problem are carried to the next higher order (see, however, Bethe (B15)).

In the second quantised formalism perturbation theory can be developed either as a time-dependent (G1, G3, S5) or time-independent theory (H1, H2). Historically, both in field theory and in the many body problem the time-dependent methods came first because they provided the most direct way of enumerating and summing the contributions from all the possible time orderings of a given set of graphs. Later, the use of convolution integrals in the complex energy plane and the Green's function or the resolvent operator led to a rapid development in the time-independent theory. At the present time, at least for the many body problem, more formal results are known through this formalism than through the time-dependent one. We have given both these methods because it is felt that at this stage of development of many body theory it will be profitable to know and investigate all possible ways of formulating the problem.

Second quantisation is introduced in the section 3 and Goldstone's proof of the linked cluster theorem (G1) is given. The theory is developed from the beginning and in detail.

A larger class of problems has been treated by Hugenholtz's time-independent method (H1, H2) in the fourth and subsequent sections. General expressions have been given for the wave function and energy of the ground state as well as metastable excited states. The characteristics of propagation of single particles through the many body medium are considered and the remarkable theorem of Hugenholtz and Van Hove (H4) concerning single particle energies is proved.

One of the assumptions in this discussion is that the system is infinitely large, but many of the results are independent of this assumption. We have tried to point out such features and also those aspects which are independent of the nature of interaction or statistics.

In the course of this development we repeatedly use the idea that with appropriate modifications in the expressions the number of terms to be actually evaluated, i.e., the diagrams which have to be 'considered', can be greatly reduced. This is the underlying idea of the so-called rearrangement or partial summation methods which are discussed in more detail in the next chapter. However, many examples already occur in the present chapter; in particular, the linked cluster theorem itself is an example of a result of partial summation.

2. THE R–S EXPANSION AND DETERMINANTAL WAVE-FUNCTIONS

In this section the first few terms of the R–S series for the case of a system of Fermions will be evaluated in an elementary way. This calculation is used to make two important points. The first is to show that the structure of terms may be very conveniently represented by means of certain diagrams, which leads to a considerable simplification in the calculational procedure. The second point concerns the systematic cancellation of certain terms in the series. These are the so-called unlinked cluster terms. The reason for this cancellation in the general case is pointed out. To illustrate various points, complete enumeration of the terms and diagrams up to third-order is given; this includes the so-called momentum conserving diagrams which are not usually given in such studies because their contribution vanishes in the limit of an infinite number of particles. This part will also be useful if one is interested in estimating the influence of such terms or the corrections to the usual second order theories.

2a. *Reduction in terms of two-particle matrix elements and clusters*

One starts with the zero order Hamiltonian in which there is no interaction among the particles.

$$K_{\text{op}} \equiv H_0 = \sum_i (T_i + U_i) \tag{2.1}$$

where T_i and U_i are the kinetic and potential energy operators of the ith particle.

If A is the number [15] of Fermions in the system then the basic wave function, Φ, of K_{op} is given by

$$\Phi = (A!)^{-\frac{1}{2}} \operatorname{Det} |\psi_1 \psi_2 \ldots \psi_A|. \tag{2.2}$$

where ψ's are the single particle wave functions satisfying the equation

$$(T_i + U_i)\psi_i = \epsilon_i \psi_i. \tag{2.3}$$

The wave function (2) describes the state in which the single particle levels $1, 2 \ldots A$ are occupied. The set $\{\Phi\}$ furnishes the p-representation of the previous chapter. K_{op} and Φ are also referred to as the model Hamiltonian and model wave function.

The actual total Hamiltonian is

$$H = \sum_i T_i + \sum_{i<j} v_{ij} = K_{\text{op}} + V. \tag{2.4}$$

From (4) and (1) we obtain

$$V = \sum_{i<j} v_{ij} - \sum_i U_i, \tag{2.5}$$

or, symmetrising the last term one could write

$$V = \sum_{i<j} \bar{v}_{ij}; \quad \bar{v}_{ij} = v_{ij} - \tfrac{1}{2}(U_i + U_j) \tag{2.5'}$$

$\bar{v}_{ij}$ is the so-called effective interaction – the perturbation in the present problem.

We wish to calculate the properties of the ground state, Ψ_0, of H. We start with the state of K_{op} expected to be closest to Ψ_0, viz., Φ_0 – the state obtained by putting the particles in the lowest available levels ψ. This configuration [16] will be called the *chosen configuration*. The states belonging to the chosen configuration will be denoted by a

[15] N is the more usual symbol for total number of particles. A is used conventionally as the number of particles in a nucleus; we use A and N interchangeably.

[16] Often the configurations are represented by merely giving the state indices on the main diagonal; thus

$$\Phi = (1\,2 \ldots i\,j \ldots A); \quad \Phi_0 = (\bar{1}\,\bar{2} \ldots \bar{i}\,\bar{j} \ldots \bar{A}).$$

bar on the subscript $\psi_{\bar{\imath}}$. That is

$$\Phi_0 = (A\,!)^{-\frac{1}{2}}\ \mathrm{Det}\ |\psi_{\bar{1}}\psi_{\bar{2}}\ \dots\ \psi_{\bar{\imath}}\psi_{\bar{\jmath}}\ \dots\ \psi_{\bar{A}}|. \tag{2.6}$$

In the general case, of course, the chosen configuration need not be the one corresponding to the lowest (model) energy.

For any operator O we shall write

$$\langle \Phi_0|O|\Phi_0\rangle \equiv \langle O\rangle \tag{2.7a}$$

and for the two-particle matrix element

$$\langle rs\,|v|\,mn\rangle = \int \psi_r^*(1)\psi_s^*(2)v_{12}\psi_m(1)\psi_n(2)\ \mathrm{d}\tau_1\,\mathrm{d}\tau_2. \tag{2.7b}$$

Now recall the R–S expression for the level shift

$$\begin{aligned}\Delta E &= \mathscr{E}_0 - E_0\\ &= \langle V\rangle + \langle V\frac{1}{a}V\rangle + \langle V\frac{1}{a}V\frac{1}{a}V\rangle + \langle V\frac{1}{a}V\frac{1}{a}V\frac{1}{a}V\rangle + \dots\\ &\qquad - \langle V\rangle\langle V\frac{1}{a}\frac{1}{a}V\rangle - \langle V\rangle\langle V\frac{1}{a}\frac{1}{a}V\frac{1}{a}V\rangle + \dots\\ &\qquad - \langle V\rangle\langle V\frac{1}{a}V\frac{1}{a}\frac{1}{a}V\rangle + \dots\\ &\qquad + \langle V\rangle^2\langle V\frac{1}{a}\frac{1}{a}\frac{1}{a}V\rangle + \dots\end{aligned} \tag{2.8}$$

where

$$\frac{1}{a} = \frac{1}{K_0 - K_{\mathrm{op}}}(1-\Lambda_0) \equiv \frac{1}{E_0 - H_0}(1-\Lambda_0).$$

To evaluate successive terms one introduces the unit operator $\Sigma_i\,|\Phi_i\rangle\,\langle\Phi_i|$ between each pair of operators in (8) and reduces the expressions in terms of two body matrix-elements like (7). The following types of matrix elements can occur:

(*a*) THE DIAGONAL ELEMENTS

$$\langle\Phi|V|\Phi\rangle = \sum_{i<j}\{\langle ij\,|\bar{v}|\,ij\rangle - \langle ij\,|\bar{v}|\,ji\rangle\} \equiv \sum_{i<j}(ij\,;ij). \tag{2.9a}$$

The double sum extends over all the states of Φ.

(*b*) THE NON-DIAGONAL ELEMENTS

(*b*) (*i*) STATES DIFFERING IN ONE EXCITATION

$$\langle\Phi|V|\Phi'\rangle = \sum_j \{\langle ij|\bar{v}|i'j\rangle - \langle ij|\bar{v}|ji'\rangle\} \equiv \sum_j (ij; i'j). \qquad (2.9b)$$

The states Φ and Φ' differ only in having $i \neq i'$ there is a single sum extending over all single particle states common to both Φ's.

(*b*) (*ii*) STATES DIFFERING IN TWO EXCITATIONS

$$\langle\Phi|V|\Phi''\rangle = \{\langle ij|\bar{v}|i'j'\rangle - \langle ij|\bar{v}|j'i'\rangle\} \equiv (ij; i'j')$$

$$i, j \neq i', j'. \text{ There is no sum.} \qquad (2.9c)$$

(*b*) (*iii*) STATES DIFFERING IN MORE THAN TWO EXCITATIONS

$$\langle\Phi|V|\Phi^{(n)}\rangle = 0. \qquad (2.9d)$$

Reducing (8) to two-body elements term by term we have [17]

First Order:

$$\langle V\rangle = \sum_{i<j} \{\langle \bar{i}\bar{j}|\bar{v}|\bar{i}\bar{j}\rangle - \langle \bar{i}\bar{j}|\bar{v}|\bar{j}\bar{i}\rangle\} \equiv \sum_{i<j} (\bar{i}\bar{j}; \bar{i}\bar{j}). \qquad (2.10)$$

Second Order:

$$\langle V \frac{1}{a} V\rangle = \sum_{n\neq 0} \langle\Phi_0|V|\Phi_n\rangle \frac{1}{E_0 - E_n} \langle\Phi_n|V|\Phi_0\rangle. \qquad (2.11)$$

Non-vanishing contribution arises in two cases: $\Phi_n = \Phi_0'$ and $\Phi_n = \Phi_0''$. The result is

$$\langle V \frac{1}{a} V\rangle = \sum_{\bar{i}\bar{j}\bar{k}r} (\epsilon_i - \epsilon_r)^{-1}[\langle \bar{i}\bar{j}|\bar{v}|r\bar{j}\rangle\langle r\bar{k}|\bar{v}|\bar{i}\bar{k}\rangle - \langle \bar{i}\bar{j}|\bar{v}|\bar{j}r\rangle\langle r\bar{k}|\bar{v}|\bar{i}\bar{k}\rangle$$

$$+ \langle \bar{i}\bar{j}|\bar{v}|\bar{j}r\rangle\langle \bar{k}r|\bar{v}|\bar{i}\bar{k}\rangle - \langle \bar{i}\bar{j}|\bar{v}|r\bar{j}\rangle\langle \bar{k}r|\bar{v}|\bar{i}\bar{k}\rangle]$$

$$+ \sum_{\bar{i}\bar{j}rs} (\epsilon_i + \epsilon_j - \epsilon_r - \epsilon_s)^{-1}[\langle \bar{i}\bar{j}|\bar{v}|rs\rangle\langle rs|\bar{v}|\bar{i}\bar{j}\rangle - \langle \bar{i}\bar{j}|\bar{v}|sr\rangle\langle rs|\bar{v}|\bar{i}\bar{j}\rangle$$

$$+ \langle \bar{i}\bar{j}|\bar{v}|sr\rangle\langle sr|\bar{v}|\bar{i}\bar{j}\rangle - \langle \bar{i}\bar{j}|\bar{v}|rs\rangle\langle sr|\bar{v}|\bar{i}\bar{j}\rangle] \qquad (2.12)$$

[17] The method followed here is slightly different from that used by Brueckner (B11, B13). He takes the matrix elements of individual members $\bar{v}_{ij}$ of the interaction term $\sum_{i<j} \bar{v}_{ij} = V$ and arranges them according to indices of $\bar{v}_{ij}$. We take the matrix element of V first and then arrange according to the state labels. In this way we always deal with state indices. In Brueckner's method one has to watch against a likely confusion between particle and state indices.

or, abbreviating with

$$(\epsilon_{\bar{i}} - \epsilon_r)^{-1} = \epsilon_{\bar{i},r} \tag{2.13a}$$

$$(\epsilon_{\bar{i}} + \epsilon_{\bar{j}} - \epsilon_r - \epsilon_s)^{-1} = \epsilon_{\bar{i}\bar{j},rs} \tag{2.13b}$$

$$\langle V \frac{1}{a} V \rangle = \sum_{\bar{i}\bar{j}\bar{k}r} \epsilon_{\bar{i},r}[(\bar{i}\bar{j}, r\bar{j})(r\bar{k}, \bar{i}\bar{k})] + \sum_{\bar{i}\bar{j},rs} \epsilon_{\bar{i}\bar{j},rs}[(\bar{i}\bar{j}, rs)(rs, \bar{i}\bar{j})]. \tag{2.14}$$

In the second term of (12) the contributions of the last two lines are equal reflecting the interchangeability of r and s. Note that due to Pauli principle or the antisymmetry of Φ's the indices in the same bra or ket in (12) cannot be equal and the unbarred subscripts range only over those values which are not present in the chosen configuration. *The summation over the single particle indices has to be suitably restricted to reproduce the effect of summing over all the distinct intermediate states* Φ_n.

A *cluster* [18] is defined as a product of a certain number, say n, of two particle matrix elements $\langle rs|\bar{v}|mn\rangle$ and $(n - 1)$ energy denominators $\epsilon_{\bar{i}\bar{j}\ldots,rs\ldots}$ between them.

A *linked cluster* is one in which every two particle matrix element has at least one index common with at least one other matrix element *and* an energy denominator.

An *unlinked cluster* is one which is not linked according to the above definition. No unlinked clusters occur in the first and second orders.

2b. *Representation by means of diagrams*

At the moment we introduce these diagrams as just a mnemonic device for remembering the terms in the perturbation expansion [18]. For this purpose we recall the convention of reading the transition matrix elements. The initial state is on the extreme right in a cluster term, the ones in between are successive intermediate states. The 'transitions', which are often visualised as virtual transitions occurring in the system, take place through the interaction $\bar{v}_{ij}$. This specifies

[18] Clusters and diagrams are here defined only for energy. More precise definitions applicable to both energy and wave function expressions are given in sections 3 and 4.

for us a sense of *time* in our matrix element. The intermediate states will be represented by drawing some directed lines. The states which occur in the matrix elements but not in the energy denominators are said to be 'not propagated' in time. The lines corresponding to these form instantaneous loops, *e.g.*, those of $\bar{k}$ in Figure 2. Strictly speaking the loop takes no time, which is of course impossible to show in a diagram. In particular it is irrelevant to put arrows on the loops. Beyond this one can establish different conventions for drawing the diagrams.

2b (i). Goldstone diagrams

The 'time' is taken to flow upwards. The interaction is represented by a horizontal dotted line to which are attached two full lines at each end representing the four indices involved. These latter, so-called particle lines (perhaps more appropriately-state lines), are given a direction to distinguish between barred (opposite to 'time' – downwards) and unbarred (in the direction of 'time' – upwards) indices. In diagrams representing total energy all particle lines finally close – there are no free lines. This latter signifies return to the chosen configuration after the final interaction.

Examples:

First order: equation (10), see Fig. 1.
Second order: equation (12). The single excitation diagrams corresponding to the first two lines are from (a) to (d). These are also called the momentum non-conserving diagrams because when the single particle states are momentum states they represent a spontaneous change in the momentum of only one particle and their contribution vanishes.

There are only two double excitation diagrams (e) and (f). The

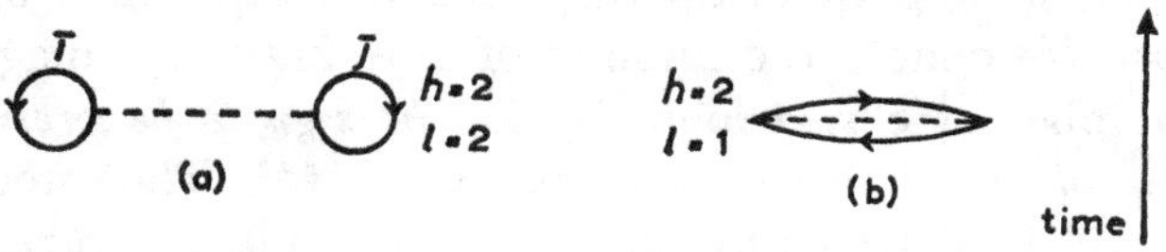

Fig. 1. Goldstone diagrams for energy in the first order: (a) direct, (b) exchange.

contributions of the third and fourth line of (12) are identical except for the interchange of the labels r and s.

Referring back to the definition of the matrix element, equation (7), one says that the element $\langle rs|\bar{v}_{12}|mn\rangle$ represents particle 2 in the

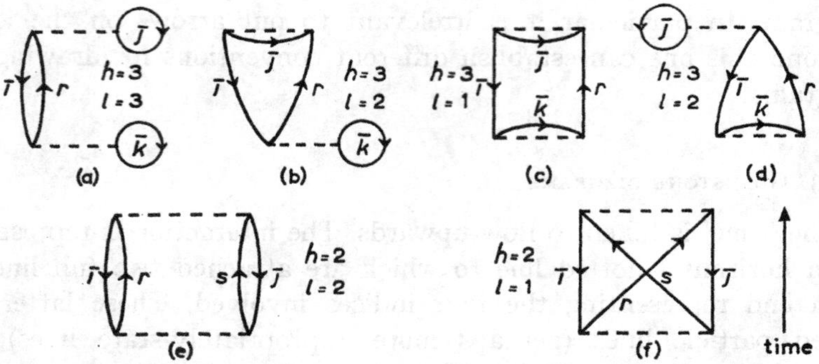

Fig. 2. Goldstone diagrams for energies in the second order. Single excitation momentum non-conserving diagrams: (a) direct, (b) and (d) single exchange (c) double exchange. Double excitation momentum conserving diagrams: (e) direct (f) exchange.

state ψ_n going to state ψ_s and particle 1 in the state ψ_m going to state ψ_r as a result of the interaction $\bar{v}_{12}$ between the particles. The same thing is abbreviated by saying that n goes into or is scattered into s and m goes into r (because of the interaction). With this much help the graph may be read from bottom to top.

The lines going backward in time indicate that that particular state of the chosen configuration is unoccupied. The lines corresponding to the occupied states of the chosen configuration do not appear in the diagrams except in the instantaneous loops. The chosen configuration is therefore called the 'vacuum' state and the lines going backward in time the 'hole lines'.

Notice that some parts of the diagrams in the Figures 1 and 2 form *closed loops*. We denote the number of hole lines in a diagram by h and the number of closed loops by l. *The sign to be prefixed to the contribution of the diagram is given by* $(-)^{h+l}$. The proof of this statement will be deferred for the time being, but its correctness can be verified for the diagrams given in the Figures 1 and 2.

2b (ii). HUGENHOLTZ DIAGRAMS

This is an alternative way of writing the diagrams. The 'time' is taken to run from right to left as in the written matrix element. The abbreviated so-called antisymmetrised form of writing the interaction, equations (10) and (14), is used and it is denoted by a dot on the diagram. As in the Goldstone diagrams the lines corresponding to the unoccupied members of the chosen configuration go backward in time, while for states not in the chosen configuration the lines go forward in time. The number of diagrams in a given order is very much less in this system than in Goldstone's. Thus we have one diagram each for first order and for double and single excitations in second order (see Fig. 3).

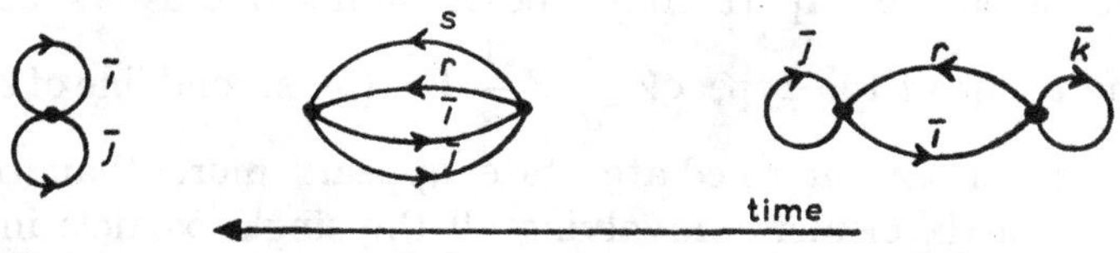

Fig. 3. Hugenholtz diagrams for energies in first and second orders. (a) First order Eq. (II. 2.10); second order (b) double excitation and (c) single excitation, second and first terms respectively on the r.h.s. of Eq. (II. 2.14).

2b (iii). BRUECKNER DIAGRAMS (B13)

Although these diagrams will not be used they are mentioned here for the sake of completeness. Chief drawback is that there is no provision for drawing hole-lines. (This is often expressed by saying – "they are not field theoretic"

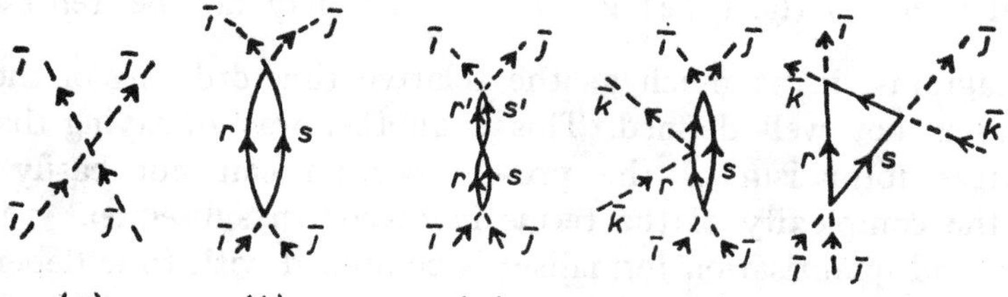

Fig. 4. Brueckner diagrams for energy in first three orders. Note that there is no provision for drawing hole lines. (a) first order, (b) second order, and (c, d, e) third order.

or that "they are not Feynman diagrams".) Hence the corrections left over after cancellation of the unlinked parts which involve hole-hole or hole-particle interaction diagrams cannot be represented by Brueckner diagrams and one has to remember them separately. The way of writing the diagrams will be clear by the examples given in Fig. 4.

Sometimes a dot is put at the vertex and the dotted lines representing states in the chosen configuration are omitted.

2c. *Cancellation of unlinked cluster terms: example of third order*

Unlinked clusters do not appear in the first and second orders. In higher orders terms of the type $\langle V\rangle \langle V \frac{1}{a} \frac{1}{a} V\rangle$, (see third and subsequent lines of equation (8)), occur which have an obviously unlinked structure. Apart from these, unlinked clusters also arise from the terms of the type $\langle V \frac{1}{a} V \frac{1}{a} V\rangle$, (see second line of equation (8)), whenever an intermediate state appears more than once. In that case matrix elements involving all the single particle indices of Φ_n occur, and the ones which are not also present in the energy denominators give rise to the unlinked clusters. We shall show below, by elementary algebraic manipulation, how the unlinked cluster contributions cancel each other in the third order. Also in this way we exhibit all the terms of the third order. The relative importance of various types of terms depends, of course, on the nature of interactions and symmetries of the system. We shall not consider this aspect for the time being.

At this point it may be noted that the contribution of the unlinked terms of the type $\langle V\rangle \langle V \frac{1}{a} \frac{1}{a} V\rangle$ may not be represented by diagrams, in as much as the relative time ordering of the two factors is not well defined. This is another way of saying that the primitive formalism of the present section can not easily cope with the complexity of the terms involved. In subsequent sections the second quantisation formalism is combined with time dependent perturbation theory and the convolution integral to carry out the formal sums in higher order. These are then used to prove, among other things, that the unlinked cluster contribution in general cancel each other exactly for all orders.

The complete third order term, equation (8), is

$$\langle V\frac{1}{a}V\frac{1}{a}V\rangle - \langle V\rangle\langle V\frac{1}{a}\frac{1}{a}V\rangle$$

$$= \sum_{n_1\neq n_2} \langle\Phi_0|V\frac{1}{a}|\Phi_{n_1}\rangle\langle\Phi_{n_1}|V|\Phi_{n_2}\rangle\langle\Phi_{n_2}|\frac{1}{a}V|\Phi_0\rangle$$

$$+ \sum_n \langle\Phi_0|V\frac{1}{a}|\Phi_n\rangle\langle\Phi_n|\frac{1}{a}V|\Phi_0\rangle[\langle\Phi_n|V|\Phi_n\rangle - \langle\Phi_0|V|\Phi_0\rangle]. \quad (2.15)$$

Consider first all the terms in the square bracket on the right hand side of (15). Two cases arise

(a) $\Phi_n = \Phi_0'$ i.e., $(\Phi_n = \ldots r\bar{j}\bar{k}\ldots;\ \Phi_0 = \ldots \bar{i}\bar{j}\bar{k}\ldots)$

$$[\ \ ] = \sum_{\bar{m}\neq i}\{(r\bar{m}, r\bar{m}) - (\bar{i}\bar{m}, \bar{i}\bar{m})\} \quad (2.16a)$$

(b) $\Phi_n = \Phi_0''$ i.e., $(\Phi_n = \ldots rs\bar{k}\ldots;\ \Phi_0 = \ldots \bar{i}\bar{j}\bar{k}\ldots)$

$$[\ \ ] = \sum_{i,\bar{j}\neq\bar{m}}[\{(r\bar{m}, r\bar{m}) + (s\bar{m}, s\bar{m})\} - \{(\bar{i}\bar{m}, \bar{i}\bar{m}) + (\bar{j}\bar{m}, \bar{j}\bar{m})\}$$

$$+ (rs, rs) - (\bar{i}\bar{j}, \bar{i}\bar{j})]. \quad (2.16b)$$

A summation over Φ_n in (a) means sum over all values of r and $\bar{i}$; in (b) it means sum over all possibilities of r, s, $\bar{i}$, $\bar{j}$. Note that all terms of the type $(\bar{k}\bar{l}, \bar{k}\bar{l})$ where $\bar{k}$ and $\bar{l}$ also occur in Φ_n, have cancelled. They would have led to unlinked diagrams.

Hence the contribution to the second term on the right hand side of Eq. (15) from the singly excited configurations in equation (16a), is given by

$$\sum_{ir}(\epsilon_{\bar{i},r})^2[\sum_{\bar{j}}(\bar{i}\bar{j}, r\bar{j})][\sum_{\bar{k}}(r\bar{k}, \bar{i}\bar{k})][\sum_{\bar{m}\neq i}(r\bar{m}, r\bar{m}) - (\bar{i}\bar{m}, \bar{i}\bar{m})]$$

$$= \sum_{i,\bar{j},\bar{k},\bar{m},r}[\epsilon_{\bar{i},r}]^2\{(\bar{i}\bar{j}, r\bar{j})(r\bar{m}, r\bar{m})(r\bar{k}, \bar{i}\bar{k}) - (\bar{i}\bar{j}, r\bar{j})(\bar{i}\bar{m}, \bar{i}\bar{m})(r\bar{k}, \bar{i}\bar{k})\}. \quad (2.17)$$

There are two terms – one direct and one exchange – in each of the parantheses hence there are sixteen triple products in the above expression. However, some of these sixteen terms are identical. Usually the diagrams are written down before the corresponding matrix elements. Here we are interested in showing the connection between the diagrams and the matrix elements. We therefore reverse the procedure and write the diagrams corresponding to the terms in (17). The eight Goldstone diagrams corresponding to the first term on the r.h.s. are shown in Fig. 5.

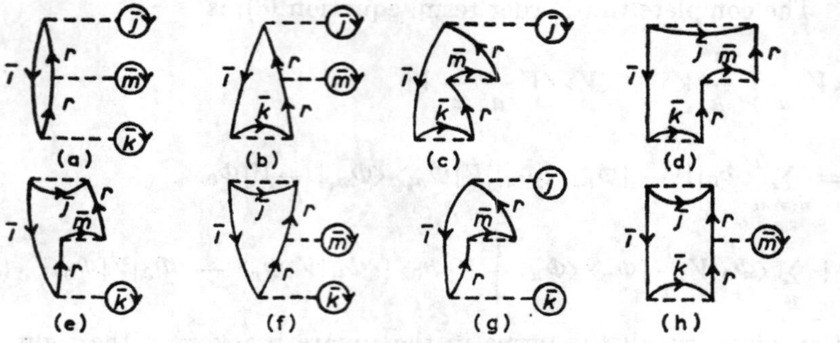

Fig. 5. Eight Goldstone diagrams of third order corresponding to the first term on right hand side of Eq. (II. 2.17). Diagrams giving equal contributions are topologically equivalent, after noting this fact and making a rule about the resulting factors equivalent diagrams may be represented by only one of their kind.

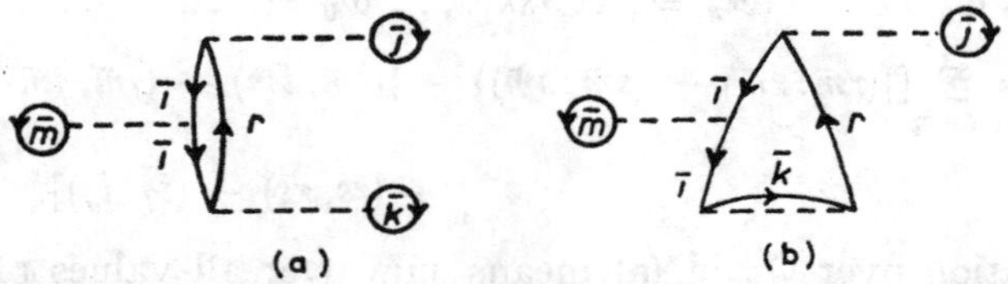

Fig. 6. Two of the Goldstone diagrams corresponding to the second term on right hand side of Eq. (II. 2.17).

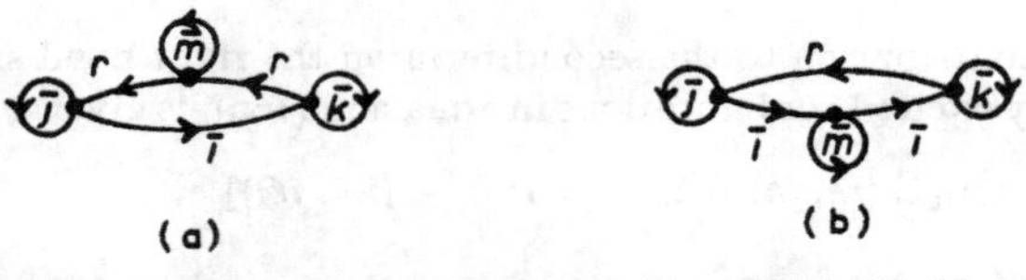

Fig. 7. Hugenholtz diagrams corresponding to Eq. (II. 2.17). Note the enormous reduction in number of diagrams (Cf., Fig. 5 and 6.)

The diagrams representing the second term on the r.h.s. of (17) differ in having only one internal particle line (r) (any line connecting two interactions is an internal line). From the structure of the terms in (17) it is seen that the (odd) internal hole line always occurs with a minus sign, thus generating the factor $(-)^h$. Also whenever there is an exchange it always implies the breakup of one loop leading to a factor $(-)^l$. Hence the total sign before a matrix

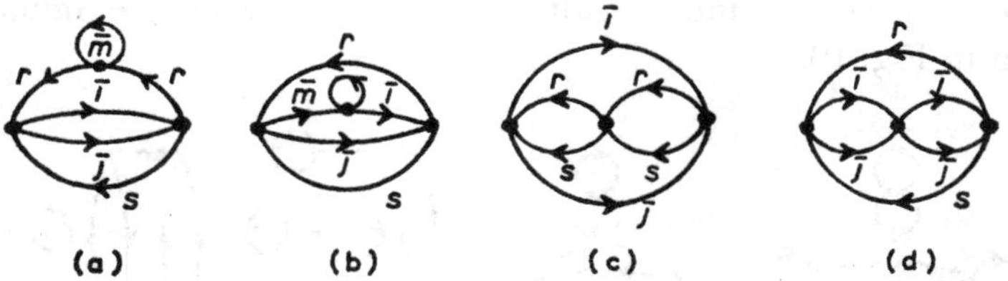

Fig. 8. Four Hugenholtz diagrams for double excitation terms in third order. Diagrams (a), (b), (c) and (d) correspond respectively to the four terms from left to right on the right hand side of (II. 2.18).

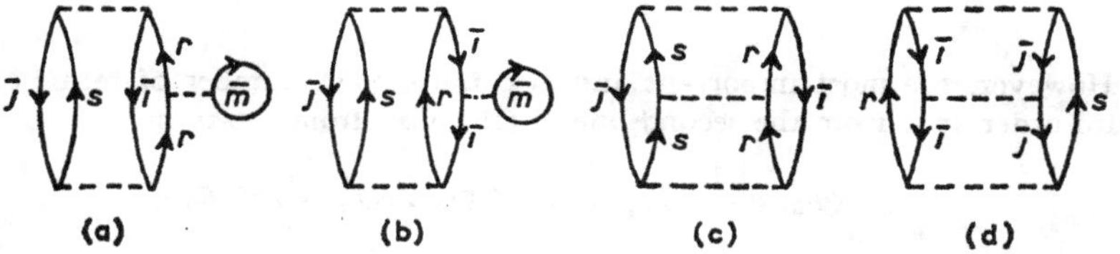

Fig. 9. Some Goldstone diagrams of double excitation in third order. (a–d) correspond respectively to the direct contribution from each of the four terms from left to right on the right hand side of (II. 2.18).

elements is $(-)^{h+l}$. This rule holds throughout, as is clear from the way the diagrams are obtained. The first two diagrams corresponding to the second term of (17) are given in Fig. 6. These diagrams are similar to those in figure 5 except that $\bar{m}$ now interacts with $\bar{i}$ instead of r, giving rise to two internal hole lines in place of one.

The Hugenholtz diagrams for these are respectively shown in the Figures 7 (a) and 7 (b).

Corresponding to the doubly excited intermediate states, equation (16), we have

$$\sum_{\bar{i}\bar{j}rs} (\epsilon_{\bar{i}\bar{j}rs})^2[(\bar{i}\bar{j}, rs)][(rs, \bar{i}\bar{j})][\text{all of r.h.s. of (16b)}]$$

$$= \sum_{\substack{\bar{m}\bar{i}\bar{j}rs \\ i,j\neq m \\ r\neq s}} (\epsilon_{\bar{i}\bar{j}rs})^2(\bar{i}\bar{j}, rs)[2(r\bar{m}, r\bar{m}) - 2(\bar{i}\bar{m}, \bar{i}\bar{m}) + (rs, rs) - (\bar{i}\bar{j}, \bar{i}\bar{j})](rs, \bar{i}\bar{j}). \quad (2.18)$$

There are four Hugenholtz diagrams corresponding to (18). (See Fig. 8.) In Goldstone's way of writing we give in Fig. 9 the first graphs corresponding to the terms in (18) with the observation that each of them generates seven more diagrams (because of exchanges) some of which may be equivalent so that the actual number of distinct diagrams may be less than thirty two.

The examples of unlinked diagrams that have been eliminated are shown in Fig. 10.

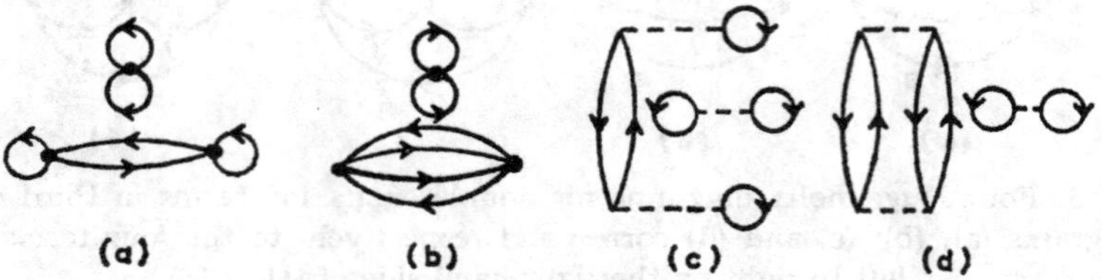

Fig. 10. Examples of unlinked diagrams that have been eliminated. (a, b) Hugenholtz; (c, d) Goldstone diagrams for the same terms.

However, the most important and also the largest number of terms of the third order arise from the second line of (15), viz., from the term

$$\sum_{n_1 \neq n_2} \langle \Phi_0 | V \frac{1}{a} | \Phi_{n_1} \rangle \langle \Phi_{n_1} | V | \Phi_{n_2} \rangle \langle \Phi_{n_2} | \frac{1}{a} V | \Phi_0 \rangle .$$

The reduction of this expression in terms of two body interactions is facilitated by classifying according to the relative structure of Φ_{n_1} and Φ_{n_2}, they can differ from each other and from the ground state at most by two excitations. Let $\Phi_0 = \ldots \bar{i}, \bar{j}, \bar{k}, \ldots$ then following cases arise:

(*i*) Φ_{n_1} and Φ_{n_2} both singly excited:

(*a*) $\Phi_{n_1} = \ldots r\bar{j}\bar{k}$ (*b*) $\Phi_{n_1} = r\bar{j}\bar{k}$

$\Phi_{n_2} = \ldots s\bar{j}\bar{k}$ $\Phi_{n_2} = \bar{i}s\bar{k}$.

The corresponding matrix elements are (summations understood) respectively

$$\epsilon_{\bar{i},r}\epsilon_{\bar{i},s}(\bar{i}\bar{j}, r\bar{j})(r\bar{k}, s\bar{k})(s\bar{l}, \bar{i}\bar{l}) \tag{2.19a}$$

$$\epsilon_{\bar{i},r}\epsilon_{\bar{j},s}(\bar{i}\bar{k}, r\bar{k})(r\bar{j}, \bar{i}s)(s\bar{l}, \bar{i}\bar{l}) \tag{2.19b}$$

with $r \neq s$, $\bar{i} \neq \bar{j}$.

(*ii*) Φ_{n_1} and Φ_{n_2} both doubly excited:

(*a*) $\Phi_{n_1} = rs\bar{k}$ (*b*) $\Phi_{n_1} = rs\bar{k}$ (*c*) $\Phi_{n_1} = rs\bar{k}$

$\Phi_{n_2} = ps\bar{k}$ $\Phi_{n_2} = pq\bar{k}$ $\Phi_{n_2} = \bar{i}st$.

The respective matrix elements are

$$\epsilon_{\bar{i}\bar{j},rs}\epsilon_{\bar{i}\bar{j},ps}\{(\bar{i}\bar{j}, rs)(rs, ps)(ps, \bar{i}\bar{j}) + (\bar{i}\bar{j}, rs)(r\bar{k}, p\bar{k})(ps, \bar{i}\bar{j})\} \quad p \neq r \tag{2.20a}$$

$$\epsilon_{\bar{i}\bar{j},rs}\epsilon_{\bar{i}\bar{j},pq}(\bar{i}\bar{j}, rs)(rs, pq)(pq, \bar{i}\bar{j}) \quad r, s \neq p, q \tag{2.20b}$$

$$\epsilon_{\bar{i}\bar{j},rs}\epsilon_{\bar{i}\bar{j},ts}(\bar{i}\bar{j}, rs)(r\bar{k}, \bar{i}t)(ts, \bar{k}\bar{j}) \quad r \neq s \neq t \tag{2.20c}$$

(*iii*) Φ_{n_1} singly excited and Φ_{n_2} doubly excited and vice-versa. If the two particle matrix elements are real the two possibilities give equal contributions. One need consider only one possibility and multiply the contribution by a factor two. The total contribution being twice the real part of the term in question. Hence we have

$$(a)\ \Phi_{n_1} = r\bar{j}\bar{k} \qquad\qquad (b)\ \Phi_{n_1} = r\bar{j}\bar{k}$$

$$\Phi_{n_2} = rs\bar{k} \qquad\qquad \Phi_{n_2} = st\bar{k}.$$

The matrix elements are

$$\epsilon_{\bar{i},r}\epsilon_{\bar{i}\bar{j},rs}\{(\bar{i}\bar{k}, r\bar{k})(r\bar{j}, rs)(rs, \bar{i}\bar{j}) + (\bar{i}\bar{k}, r\bar{k})(\bar{l}\bar{j}, \bar{l}s)(rs, \bar{i}\bar{j})\} \tag{2.21a}$$

$$\epsilon_{\bar{i},r}\epsilon_{\bar{i}\bar{j},st}(\bar{i}\bar{k}, r\bar{k})(r\bar{j}, st)(st, \bar{i}\bar{j}) \quad t \neq r \neq s. \tag{2.21b}$$

If the restriction $t \neq r$ is dropped in (21b) it will include the first term of (21a). If the restriction $r, s \neq p, q$ is dropped in (20b) it will include both the first term of (20a) and the contribution of Fig. 8c. If the restriction $r \neq s$ is dropped in (19a) it will include the first term of (17).

All the distinct Hugenholtz diagrams of third order arising in this way are collected in Fig. 11. Some state labels have been changed.

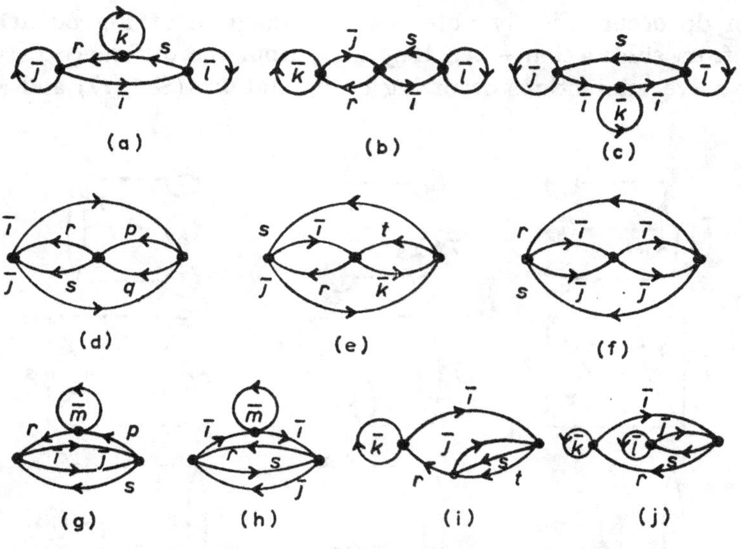

Fig. 11. All distinct, linked Hugenholtz diagrams for third order energy contribution. (a) (19a); (b) (19b); (c) second term of (17); (d) first term of (20a and b) and Fig. 8c; (e) (20c); (f) Fig. 8d; (g) Fig. 8a and second term of (20a); (h) Fig. 8b; (i) (21b) and first term of (21a); (j) second term of (21a).

A *passive* hole line (e.g. ◯--- or ⇄) represents an interaction with a particle occupying a state in the chosen configuration. In contrast a *propagating* hole line represents the absence of a particle in a state of chosen configuration. Hence interaction with a propagating hole line is not always possible. This shows that there would *not* be complete symmetry between the hole and particle lines in the diagrams. Thus there are some diagrams involving hole interactions which *do not* occur. (Examples are shown in Fig. 12.)

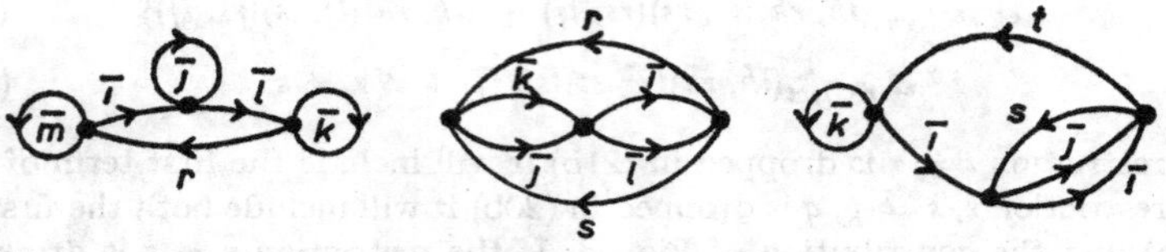

Fig. 12. Examples of linked diagrams that do not occur, i.e., have no counter part in perturbation theory expansion.

However, similar diagrams involving an intermediate particle interaction do occur. Finally note that the diagram of Fig. 8d and 8b arise from subtraction – i.e., they arise from the difference between the passive interactions occurring in Φ_n and Φ_0. (See (17) and (18).)

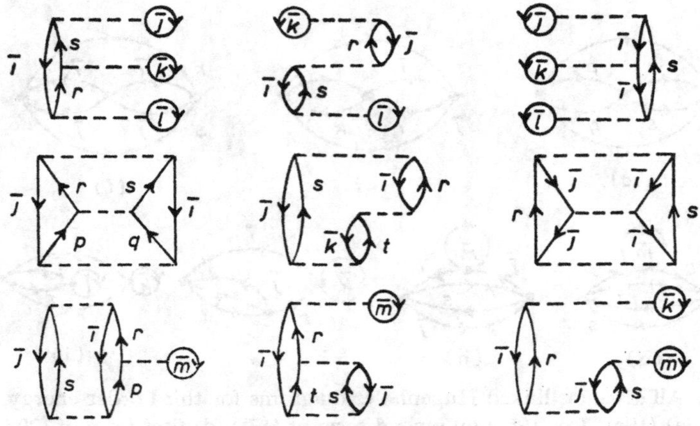

Fig. 13. Goldstone diagrams corresponding to Fig. 11. Only leading diagrams are given (all possible distinct exchanges are to be made on each one).

In the foregoing we have tried to show how one may represent matrix elements by means of diagrams. In particular we have learned to recognise three distinct classes of diagrams: the unlinked, linked and momentum-non-conserving diagrams. We also learned that one needs to consider only distinct diagrams. It was noticed that care must be exercised in considering hole-particle or hole-hole interaction because of the lack of symmetry between holes and particles in the present theory.

In remaining sections of this chapter the rules of calculating with diagrams will be derived for the general case by introducing second quantisation formalism (the so-called field theoretic methods). The power of this device will be seen in various general theorems that follow from its use.

3. TIME DEPENDENT THEORY

3a. *Second quantisation*

We will here briefly recapitulate the main features of this formalism to fix the notation and emphasise some points of current interest.

Let the symbol $\boldsymbol{x}$ stand for a set $(\boldsymbol{x}_i)$ of arbitrary number of particle coordinates $\boldsymbol{x}_i$. Similarly let k and n be the symbols for the sets (k_i) and (n_{k_i}) respectively where k_i is the ith particle state denoted in the last section simply by i and n_{k_i} is the number of particles occupying the state k_i. The symbol n_i will be used interchangeably with n_{k_i}.

The normalised many particle wave function for an N particle system may be written as

$$\langle \boldsymbol{x}|k\rangle \to \Phi_k(\boldsymbol{x}) = \sqrt{\frac{N!}{n_1! \ldots n_N!}}\, \Lambda[\psi_{k_1}(\boldsymbol{x}_1) \ldots \psi_{k_N}(\boldsymbol{x}_N)]. \tag{3.1}$$

The operator Λ generates the correct statistical properties of the wave function. For Fermions Λ is the so-called anti-symmetriser: $\Lambda = \mathscr{A} = (N!)^{-1} \sum_P \theta_P P$, where P is the permutation operator and $\theta_P = \pm 1$ depending on the odd-even character of the permutation. For Bosons the wave function is symmetric so that $\Lambda = \mathscr{S} = (N!)^{-1} \sum_P P$. $\mathscr{S}$ is called the symmetriser.

$\Phi_k(\boldsymbol{x})$ is the representative of the state $|k\rangle$ in the coordinate representation. An alternative representation of the same state is obtained by specifying the number n_i of particles occupying the states k_i, i.e.,

$$\langle n|k\rangle \to |n_{k_1} \ldots n_{k_N} \ldots\rangle. \tag{3.2}$$

The advantage in the latter representation is that systems containing arbitrary numbers of particles may be treated on the same footing. To restrict to a system with fixed number of particles one has to require $\sum_i n_i = N$. Transition from one to the other representation is provided by means of the matrix

$$\langle n|\boldsymbol{x}\rangle = \sum_k \langle n|k\rangle \langle k|\boldsymbol{x}\rangle = \sum_k \Phi_k^*(\boldsymbol{x})|n_1 \ldots n_i \ldots\rangle. \tag{3.3}$$

In the occupation number representation (2) let us introduce creation and annihilation operators η_i^* and η_i for the particles in the ith state by means of the relations (*indices k_i and i will be used interchangeably*).

$$\eta_i^*|n_1 \ldots n_i \ldots\rangle = \sqrt{n_i + 1}\,|n_1 \ldots n_i + 1 \ldots\rangle \tag{3.4a}$$

$$\eta_i|n_1 \ldots n_i \ldots\rangle = \sqrt{n_i}\,|n_1 \ldots n_i - 1 \ldots\rangle. \tag{3.4b}$$

It follows that the number of particles can be obtained in this representation by taking the expectation value of the operator.

$$N_{\text{op}} = \sum_{i=0}^{\infty} \eta_i^* \eta_i. \tag{3.5}$$

Also it follows that all states, with arbitrary numbers of particles, can be generated by operating on the lowest state, the vacuum, with an appropriate number of creation operators. Thus

$$|n_{k_1} \ldots n_{k_N} \ldots\rangle = \eta_{k_1}^* \ldots \eta_{k_N}^* \ldots |0_A\rangle \tag{3.2'}$$

where $|0_A\rangle$ is the absolute vacuum state in which there are no particles. This is the state of lowest energy in the absolute sense. Later on we shall introduce the lowest energy state of an N-particle system and that also will be called vacuum state in conformity with the usual convention. No confusion should arise.

Indeed any state can be obtained from any other state by operating with an appropriate number of creation or annihilation operators. In particular one can show (T7) that any N-particle wave function $|\Phi\rangle$

which is not orthogonal to the wave function $|\Phi_0\rangle$ given by

$$|\Phi_0\rangle = (\prod_{i=1}^{N} \eta_i^*)|0_A\rangle$$

can be written in the form

$$|\Phi\rangle = [\prod_{i=1}^{N} \prod_{m=N+1}^{\infty} (1 + C_{mi}\eta_m^*\eta_i)]|\Phi_0\rangle$$

where the coefficients C_{mi} are uniquely determined .Conversely any wave function written as above is a proper N-particle wave function which takes the statistics of particles automatically into account.

If there is only one particle in the field then the wave function in this representation is

$$\langle n|k_i\rangle = \eta_i^*|0_A\rangle. \tag{3.6}$$

The wave function of the single-particle state in coordinate representation is $\psi_{k_i}(x_1)$.Hence from (3) and (6) the transformation matrix $\psi^*(x_1)$ is

$$\langle n|x_1\rangle = \sum_{k_1} \psi_{k_1}^*(x_1)\eta_{k_1}^*|0_A\rangle = \psi^*(x_1)|0_A\rangle. \tag{3.7}$$

Similarly,

$$\langle x_1|n\rangle = \sum_{k_1} \langle 0_A|\eta_{k_1}\psi_{k_1}(x_1) = \langle 0_A|\psi(x_1). \tag{3.7a}$$

The number operator is then

$$N_{op} = \int \psi^*(x_i)\psi(x_i)\, dx_i \tag{3.8}$$

where $\int dx_i$ represents summation and integration over the full domain of the coordinates x_i which may include spin and isotopic spin coordinates apart from the usual space coordinates.

If the particles obey Fermi statistics the operators must satisfy the (anti-) commutation relations

$$\{\eta_i, \eta_j^*\} = \delta_{ij};\ \{\eta_i, \eta_j\} = \{\eta_i^*, \eta_j^*\} = 0 \tag{3.9}$$

where

$$\{A, B\} = AB + BA.$$

For Bose statistics one has the commutation relations instead

$$[\eta_i, \eta_j^*] = \delta_{ij};\ [\eta_i, \eta_j] = [\eta_i^*, \eta_j^*] = 0 \tag{3.10}$$

where

$$[A, B] = AB - BA.$$

These relations can be proved from (4). It follows using the definition (7) of ψ that

$$\{\psi(x_1), \psi^*(x_2)\} = \delta(x_1 - x_2)$$
$$\{\psi(x_1), \psi(x_2)\} = \{\psi^*(x_1), \psi^*(x_2)\} = 0. \tag{3.11}$$

For states containing N particles one has

$$\langle n|x_1 \ldots x_N\rangle = \sum_k \langle n|k\rangle \langle k|x_1 \ldots x_N\rangle$$

$$= (N!)^{-1} \sum_{k_1 \ldots k_N} \Phi^*_{k_1 \ldots k_N}(x_1 \ldots x_N)\eta_1^* \ldots \eta_N^*|0_A\rangle$$

$$= (N!)^{-\frac{1}{2}}\psi^*(x_1) \ldots \psi^*(x_N)\ |0_A\rangle. \tag{3.13}$$

In the last step use has been made of commutation relations (10) and (11) and the form of Φ.

The wave function $\Phi_k(x)$ may now be obtained from the state vector in occupation number representation from (1) and (7)

$$\langle x|k\rangle = \Phi_{k_1 \ldots k_N}(x_1 \ldots x_N) = \sum_n \langle x_1 \ldots x_N|n\rangle \langle n|k_1 \ldots k_N\rangle$$

$$= (N!)^{-\frac{1}{2}}\langle 0_A|\psi(x_N) \ldots \psi(x_1)|n_{k_1} \ldots n_{k_N}\rangle. \tag{3.14}$$

The inverse of this transformation is clearly

$$|n_{k_1} \ldots n_{k_N}\rangle$$
$$= (N!)^{-\frac{1}{2}} \int \psi^*(x_1) \ldots \psi^*(x_N)\Phi_{k_1 \ldots k_N}(x_1 \ldots x_N)\, dx_1 \ldots dx_N|0_A\rangle. \tag{3.15}$$

For Fermions a convention is necessary to fix the sign of the wave function, usually one labels the states k_i in order of increasing energy. Thus $i < j$ if $\epsilon_i < \epsilon_j$.

To obtain the form of the operators in the new representation consider a symmetric operator $O(x_1 \ldots x_N)$ of N coordinates.

$$O(x_1 \ldots x_N)\Phi_{k_1 \ldots k_N}(x_1 \ldots x_N) = \langle x|O|k\rangle$$
$$= (N!)^{-1} \sum_{k_1' \ldots k_N'} \langle k_1' \ldots k_N'|O|k_1 \ldots k_N\rangle_a \Phi_{k_1' \ldots k_N'}(x_1 \ldots x_N). \tag{3.16}$$

Use the transformation (14) to change representation on both sides

$$O|n_{k_1} \ldots n_{k_N}\rangle = (N!)^{-1} \sum_{k_1' \ldots k_N'} \langle k_1' \ldots k_N'|O|k_1 \ldots k_N\rangle_a |n_{k_1'} \ldots n_{k_N'}\rangle \tag{3.17}$$

or

$$O|n_1 \ldots n_N\rangle$$
$$= (N!)^{-1} \sum_{k_1' \ldots k_N'} \langle k_1' \ldots k_N'|O|k_1 \ldots k_N\rangle_a \eta_{1'}^* \ldots \eta_{N'}^* \eta_N \ldots \eta_1 \eta_1^* \ldots \eta_N^*|0_A\rangle.$$

So that finally

$$O = \frac{1}{(N!)^2} \sum_{k_1' \ldots k_N';\, k_1 \ldots k_N} \langle k_1' \ldots k_N'|O|k_1 \ldots k_N\rangle_a \eta^*_{k_1'} \ldots \eta^*_{k_N'}\eta_{k_N} \ldots \eta_{k_1}. \tag{3.18}$$

The factor $(1/N!)^2$ appears because we have introduced summation over indices $k_1 \dots k_N$ for sake of symmetry. The operator O gives zero while acting on any states containing less than N particles.

Special Cases: Recall from (16) that

$$\langle k_1' \dots k_N' | O | k_1 \dots k_N \rangle_a$$

$$= \int \Phi^*_{k_1' \dots k_N'}(x_1 \dots x_N) O(x_1 \dots x_N) \Phi_{k_1 \dots k_N}(x_1 \dots x_N) \mathrm{d}x_1 \dots \mathrm{d}x_N$$

1) $$O(x_1 \dots x_N) = 1$$

$$\langle k_1' \dots k_N' | O | k_1 \dots k_N \rangle_a = N!\, \Lambda \{ \delta_{k_1' k_1} \dots \delta_{k_N' k_N} \}$$

where the permutation $P = P_k$ now permutes the unprimed indices and it can be seen on using appropriate commutation relations that $O = 1$ in occupation number representation also, as one expects.

2) $$O(x_1 \dots x_N) = \sum_{i=1}^{N} O(x_i)$$

$$\langle k_1' \dots k_N' | O | k_1 \dots k_N \rangle_a = N!\, \Lambda \{ \delta_{k_1' k_1} \dots \langle k_i' | O | k_i \rangle \delta_{k_N' k_N} \}.$$

So that

$$O = \sum_{k_1' k_1} \langle k_1' | O | k_1 \rangle \eta^*_{k_1'} \eta_{k_1} \tag{3.19}$$

3) $$O = \tfrac{1}{2} \sum_{i,j} O(x_i, x_j)$$

$$\langle k_1' \dots k_N' | O | k_1 \dots k_N \rangle_a = N!\, \Lambda \{ \delta_{k_1' k_1} \dots \langle k_i' k_j' | O | k_i k_j \rangle \dots \delta_{k_N' k_N} \}$$

so that

$$O = \tfrac{1}{4} \sum_{k_1' k_2' k_1 k_2} \langle k_1' k_2' | O | k_1 k_2 \rangle_a \eta^*_{k_1'} \eta^*_{k_2'} \eta_{k_2} \eta_{k_1}. \tag{3.20}$$

It is seen that in the above forms for one and two particle operators there is no reference to the number N of the particles in the system. In fact these forms describe one and two particle interactions in a system with *an arbitrary number* of particles. This is ani mportant point which shows the superiority of this representation. The fact that we had to start with explicit reference to an N-particle system may be likened to the normalisation to a finite volume in problems connected with plane waves.

We give finally the forms for operators of interest.

$$H = H_0 + V; \quad H_0 = \sum_{i=1}^{N} (T_i + U_i); \quad V = \sum_{i<j} v(i, j)$$

$$H_0 = \sum_i \epsilon_i \eta_i^* \eta_i; \quad (T_i + U_i)\psi_i = \epsilon_i \psi_i \tag{3.21}$$

$$T = \sum_i T_i = \frac{\hbar^2}{2m} \sum_{ij} \left(\int \nabla \psi_i^* \cdot \nabla \psi_j \, \mathrm{d}x \right) \eta_i^* \eta_j$$

$$= \frac{\hbar^2}{2m} \int \nabla \psi^*(x) \cdot \nabla \psi(x) \, \mathrm{d}x \tag{3.22}$$

$$U = \sum_i U_i = \sum_{i,j} \left(\int \psi_i^*(x) U(x) \psi_j(x) \, \mathrm{d}x \right) \eta_i^* \eta_j$$

$$= \int \psi^*(x) U(x) \psi(x) \, \mathrm{d}x \tag{3.23}$$

$$V = \tfrac{1}{2} \sum_{ijkl} \langle ij|v|kl \rangle \eta_i^* \eta_j^* \eta_k \eta_l = \int \psi^*(x_1) \psi^*(x_2) v(x_1, x_2) \psi(x_1) \psi(x_2) \, \mathrm{d}x_1 \, \mathrm{d}x_2$$

$$= \tfrac{1}{4} \sum_{ijkl} (ij, kl) \eta_i^* \eta_j^* \eta_k \eta_l. \tag{3.24}$$

The factors in last expressions are sometimes omitted and suitable restrictions are placed on summations. We have

$$(ij, kl) = (k_i k_j |v| k_k k_l)_a$$

where the related quantities have been defined in Eqs. (2.7) and (2.9).

Strictly speaking the present formalism breaks down for singular potentials like those of particles with impenetrable cores. In the following we proceed as if this difficulty did not exist. This is justified because we can always imagine that we work with a well behaved potential when making rearrangements of perturbation expressions etc. and only in the last step, when the reaction matrix has been introduced (Chapters III and IV), we let the potential become infinite. However, the formalism itself may be modified to take into account the singular nature of interactions. This involves redefinition of the creation and annihilation operators and a modification of the commutation rules (S1, S15).

3b. *Time dependent perturbation theory* (*G1*, *G3*)

The state vector of the chosen configuration Φ_0 will hereafter be referred to as the vacuum state. This is a state of the system of non-interacting or free particles – an eigenstate of H_0. The state vector of the total interacting system Ψ_0 an (eigenstate of H) corresponding to Φ_0 is hereafter called the true vacuum. The aim is to express Ψ_0 in terms of Φ_0 and the operators available.

The Schrödinger equation for the total system ($\hbar = 1$) is

$$\mathrm{i}\frac{\partial}{\partial t}\Psi_{\mathrm{S}}(t) = H\Psi_{\mathrm{S}}(t) = [H_0 + V]\Psi_{\mathrm{S}}(t). \tag{3.25}$$

We transform to an interaction representation such that it reduces to the Heisenberg representation at $t = 0$ by means of the unitary transformation

$$\Psi(t) = \exp(\mathrm{i}H_0 t)\Psi_{\mathrm{S}}(t). \tag{3.26}$$

The reason for this particular choice is that we want our state vector $\Psi(0)$ to be the same as in the Heisenberg representation, as is customary.

The time-dependence of the Heisenberg operators is of the form

$$O_{\mathrm{H}}(t) = \exp(\mathrm{i}Ht)O(0)\exp(-\mathrm{i}Ht). \tag{3.27a}$$

Consequently the time-dependence of the same operator in this particular interaction representation will be of the form

$$O(t) = U(t, 0)O_{\mathrm{H}}(t)U^{-1}(t, 0) \tag{3.27b}$$

where

$$U(t, 0) = \exp(\mathrm{i}H_0 t)\exp(-\mathrm{i}Ht) \tag{3.28}$$

is the unitary operator that takes the Heisenberg operators at time t into the interaction operators at time t in such a way that at $t = 0$ both are the same. Thus the state vector in the Heisenberg representation is related to that in the interaction representation by

$$\Psi(t) = U(t, 0)\Psi_{\mathrm{H}}(0). \tag{3.26'}$$

It follows that $U(t, 0)$ is completely determined by the differential equation and boundary condition

$$\mathrm{i}\frac{\mathrm{d}}{\mathrm{d}t}U(t, 0) = V(t)U(t, 0);\; U(0, 0) = 1 \tag{3.29}$$

where

$$V(t) = \mathrm{e}^{\mathrm{i}H_0 t}\, V\, \mathrm{e}^{-\mathrm{i}H_0 t}. \tag{3.30}$$

In other words, $V(t)$ may be obtained from V of (24) by replacing the operators η_n^* and η_n by their interaction representation forms $\eta_n^* \, e^{i\epsilon_n t}$ and $\eta_n \, e^{-i\epsilon_n t}$ respectively.

We wish to solve these equations so as to express Ψ_0 as a function of Φ_0. For this purpose it is convenient to imagine that initially, in the infinite past, there was no interaction among the particles and we start from there with a state Φ_0 of the free particle Hamiltonian. Then interaction is switched on infinitely slowly such that at the time $t = 0$ it attains its full value. It is then claimed that Φ_0 has now been transformed in to Ψ_0. The process of switching the interaction on and off is prescribed by defining the operator $U_\alpha(t, 0)$ by means of an equation analogous to (29).

$$\mathrm{i}\, \mathrm{d}U_\alpha(t, 0)/\mathrm{d}t = V(t)U_\alpha(t, 0) \exp(\alpha t); \quad U_\alpha(0, 0) = 1 \tag{3.31}$$

where α is a very small positive number. As α tends to zero (31) goes into (29) and so must their solutions. The $e^{\alpha t}$ factor may be thought to be associated [19] with $V(t)$. For finite α we may write

$$\Psi_0(\alpha) = U_\alpha^{-1}(-\infty, 0)\Phi_0 = U_\alpha(0, -\infty)\Phi_0. \tag{3.32}$$

Successive iterations of (31) give for U_α

$$U_\alpha(0, -\infty) = \sum_{n=0}^{\infty} (-\mathrm{i})^n \int_{t_1 > t_2 \ldots > t_n} V(t_1)V(t_2) \ldots V(t_n) \exp[\alpha(t_1 + t_2 + \ldots t_n)] \times \mathrm{d}t_1 \ldots \mathrm{d}t_n \tag{3.32a}$$

or

$$U_\alpha(0, -\infty) = \sum_{n=0}^{\infty} \frac{(-\mathrm{i})^n}{n!} \int_{-\infty}^{0} \mathrm{d}t_1 \ldots \mathrm{d}t_n \exp[\alpha(t_1 + t_2 + \ldots + t_n)] \times P[V(t_1)V(t_2) \ldots V(t_n)]. \tag{3.32b}$$

In (32b) the limits on all the time integrations extend from $-\infty$ to 0 independently and $P(\ldots)$ is the time ordered product (S5) of the interaction operators. That is,

$$P[V(t_1)V(t_2) \ldots V(t_n)] = V(t_i)V(t_j) \ldots V(t_n) \tag{3.33}$$

with

$$t_i > t_j > \ldots > t_n.$$

[19] In scattering problems one needs to express the final states also as eigenvectors of free particle Hamiltonian. This requires the interaction to be switched off also in infinite future ($t = \infty$), in that case the factor should be written as $\exp(-\alpha|t|)$.

If E_0 is the eigenvalue of H_0 corresponding to Φ_0 then we have the identity,

$$(H_0 - E_0)\Psi_0(\alpha) = [H_0, U_\alpha(0, -\infty)]\Phi_0. \tag{3.34}$$

Using the integral representation (32b) of U_α and the relations

$$[a, \prod_{i=1}^{n} b] = \sum_{i=0}^{n} b_1 \ldots b_i[a, b_{i+1}]b_{i+2} \ldots b_n$$

$$= \sum_{i=0}^{n} (\prod_{j=1}^{i} b_j)[a, b_{i+1}] \prod_{k=i+2}^{n} b_k; \quad (b_0 = 1)$$

and

$$[H_0, V] = \frac{\partial V}{\partial t}.$$

We have for the right hand side of (34)

$$\frac{1}{i} \sum_{n=1}^{\infty} \frac{(-i)^n}{n!} \int_{-\infty}^{0} dt_1 \ldots dt_n \exp[\alpha(t_1 + \ldots + t_n)] \sum_{l=1}^{n} \frac{\partial}{\partial t_l} P[V(t_1) \ldots V(t_n)]\Phi_0$$

$$= -\sum_{n=1}^{\infty} \frac{(-i)^{n-1}}{(n-1)!} \int_{-\infty}^{0} dt_1 \ldots dt_n \exp[\alpha(t_1 + \ldots + t_n)] \frac{\partial}{\partial t_1} P[V(t_1) \ldots V(t_n)]\Phi_0.$$

Integrating partially w.r.t. t_1 this becomes

$$= -V(0)U_\alpha(0, -\infty)\Phi_0$$

$$+ \alpha \sum_{n=1}^{\infty} \frac{(-i)^{n-1}}{(n-1)!} \int_{-\infty}^{0} dt_1 \ldots dt_n \exp[\alpha(t_1 + \ldots + t_n)] P[V(t_1) \ldots V(t_n)]\Phi_0.$$

Or, associating the usual parameter λ with V (i.e., replacing V by λV) one has,

$$[H - E_0]\Psi_0(\alpha) = i\alpha\lambda \frac{\partial}{\partial \lambda} \Psi_0(\alpha). \tag{3.35}$$

Note that the total Hamiltonian H occurs on the left hand side and λ here has the significance of a coupling constant.

Let us now define

$$X_\alpha = \Psi_0(\alpha)/\langle\Phi_0|\Psi_0(\alpha)\rangle. \tag{3.36}$$

Taking logarithms, differentiating w.r.t. λ and using (35) one has

$$\left[H - E_0 - i\alpha\lambda \frac{\partial}{\partial \lambda}\right] X_\alpha = X_\alpha \left[i\alpha\lambda \frac{\partial}{\partial \lambda} \log \langle\Phi_0|\Psi_0(\alpha)\rangle\right]. \tag{3.37}$$

On noting that $\langle\Phi_0|X_\alpha\rangle = 1$ we have, from (37)

$$\langle\Phi_0|H - E_0|X_\alpha\rangle = \mathrm{i}\alpha\lambda\frac{\partial}{\partial\lambda}\log\langle\Phi_0|\Psi_0(\alpha)\rangle.$$

So that (37) may be written as

$$(H - E_0)X_\alpha = \langle\Phi_0|H - E_0|X_\alpha\rangle X_\alpha + \mathrm{i}\alpha\lambda\frac{\partial}{\partial\lambda}X_\alpha. \tag{3.38}$$

Since in the limit of $\alpha \to 0$ the last term of (38) vanishes

$$(H - E_0)\lim X_\alpha = \langle\Phi_0|H - E_0|\lim X_\alpha\rangle \lim X_\alpha. \tag{3.39}$$

If Ψ_0 diagonalises H and ΔE is the corresponding level shift we have

$$H\Psi_0 = (E_0 + \Delta E)\Psi_0; \quad H_0\Phi_0 = E_0\Phi_0.$$

On comparing this with (39) we have

$$\Psi_0 = \lim_{\alpha\to 0}\frac{U_\alpha(0, -\infty)\Phi_0}{\langle\Phi_0|U_\alpha(0, -\infty)|\Phi_0\rangle} \tag{3.40}$$

and

$$\Delta E = \langle\Phi_0|V|\Psi_0\rangle = \lim_{\alpha\to 0}\frac{\langle\Phi_0|VU_\alpha(0, -\infty)|\Phi_0\rangle}{\langle\Phi_0|U_\alpha(0, -\infty)|\Phi_0\rangle}. \tag{3.41}$$

3c. *Linked cluster theorem* (GI)

It will be shown that the above expressions for Ψ_0 and ΔE contain only contributions from linked diagrams.

An unlinked part in a given diagram for $U_\alpha\Phi_0$ has no external lines and is not connected to the rest of the diagram by any interaction. Diagrams with no unlinked parts are called linked diagrams.

Consider all the distinct diagrams that arise when the expression for $U_\alpha\Phi_0$ is expanded by introducing the explicit form of V in terms of the creation and annihilation operators. Some of these diagrams, which correspond to the individual terms of the sum, will contain unlinked parts. That is, the subscripts on the creation and annihilation operators divide in two or more groups such that they have no common time labels. Consider the case when there are only two such sets and assign the times $0 > t_1 > \ldots > t_n$ and $0 > t'_1 > \ldots > t'_n$ respectively to the interactions involved. In virtue of the fact that no indices are common among the two groups and that there is an even number of Fermion (anti-commuting) operators in each V the value of the time integration is independent of the relative time ordering of the two sets. Actually it includes the contributions of all possible relative time orderings of the two diagrams and may be expressed as a product of two parts. It then follows from (32) that the total, $U_\alpha \Phi_0$, may be expressed as a product of two terms: one arising from the linked clusters and the other arising from the unlinked parts. Sum of the contributions of the unlinked parts is exactly equal to $\langle \Phi_0 | U_\alpha | \Phi_0 \rangle$. Hence Ψ_0 is obtained by considering only the linked graphs of $U_\alpha \Phi_0$.

Two points are essential for the derivation of the above result (a) that the number of Fermion operators in $V(t)$ must be even (b) that in labelling the intermediate lines of the diagrams no account is taken of the Pauli principle. The second requirement means that the diagrams like those in Fig. 41 occur in the totality of diagrams. This is essential for the separation of the unlinked cluster contribution. Before this separation the contributions of the Figures 14a and b will

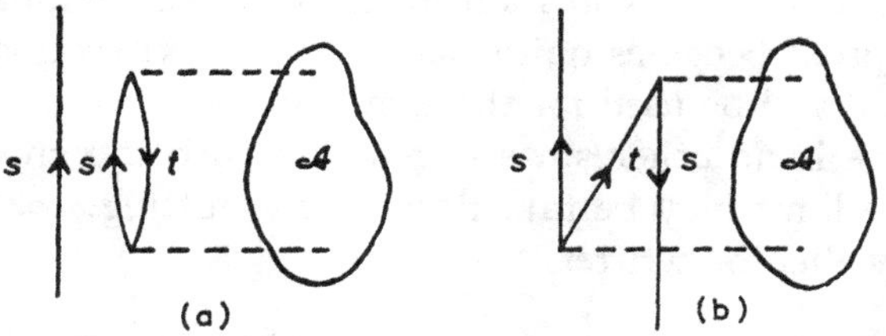

Fig. 14. Diagrams (a) and (b) violate the Pauli principle. Originally (a) cancels the contribution of diagram (b). After separation of unlinked parts (b) remains, a contribution from intermediate states violating the Pauli principle. $\mathscr{A}$ is an arbitrary subdiagram having one vertex at the top and one at the bottom.

cancel each other. In other words, whether or not unantisymmetrised intermediate states are present in the calculation makes no difference if we are not separating the unlinked parts. If we want to separate the unlinked parts, we must include the intermediate states which violate the Pauli principle: otherwise the separation would not be possible. The importance of such contributions has been emphasised by Katz (K5). This is perhaps the most important difference between the present treatment and that of the previous section where only the antisymmetric intermediate states were used. However, the role of this restriction had not become clear in the example of that section. Nonetheless, it is true that one may prove the linked cluster theorem by purely algebraic methods. On performing time integrations one has

$$\Psi_0 = \lim_{\alpha \to 0} \sum_{n=0}^{\infty} [(E_0 - H_0 + \mathrm{i}n\alpha)^{-1} V(0)(E_0 - H_0 + \mathrm{i}(n-1)\alpha)^{-1} V(0) \ldots$$
$$\ldots V(0)(E_0 - H_0 + \mathrm{i}2\alpha)^{-1} V(0)(E_0 - H_0 + \mathrm{i}\alpha)^{-1} V(0)]\Phi_0. \quad (3.42)$$

By repeated use of the identity

$$A|\Phi_0\rangle = \{\langle A\rangle + (1 - \Lambda_0)A\}|\Phi_0\rangle \quad (3.43)$$

which is valid for any operator A, one may show that Ψ_0 and ΔE are given by the familiar Rayleigh-Schrödinger series. To prove the linked cluster theorem one has to reduce (42) in terms of diagrams before using the identity (43) and taking the limit. Difficulty arises in keeping careful track of *different possible relative orderings* of the various parts. The point about using the time dependent method is that it gives a concise expression for all possible time orderings and the general proof becomes quite easy. In the next section the convolution integral will be used for the same purpose.

When there is no other state degenerate with the chosen configuration Φ_0 the limit may be taken by simply putting $\alpha = 0$. The final result [20] may then be written

$$\Psi_0 = \sum_{\mathrm{L}} \left(\frac{1}{E_0 - H_0} V\right)^n \Phi_0 \quad (3.44)$$

$$\Delta E = \sum_{\mathrm{L}} \langle\Phi_0| V \left(\frac{1}{E_0 - H_0} V\right)^n |\Phi_0\rangle. \quad (3.45)$$

[20] Kohn and Luttinger (K3) and Luttinger and Ward (L1) have pointed out that the formulae obtained in this way starting from a spherically symmetric chosen configuration are valid, in general, only for systems which are finally spherically symmetric (see also Wentzel (W6)). This result is similar to the cases discussed earlier by Rellich (R2) and is connected with the apparent breakdown of ordinary perturbation theory when the term independent of λ is missing in the expansion. Klein (K4) has shown how one may choose a

The subscript L implies that the summation extends only over linked diagrams.

3c (i). RULES FOR GOLDSTONE DIAGRAMS

These rules are implicit in the foregoing but we collect them here for ease of reference. In any order draw all possible distinct diagrams. Label the hole and particle lines without regard to the Pauli principle. To calculate the contribution of a given diagram write for each dotted line (the interaction vertex) the appropriate v-matrix element and the $e^{\pm i\epsilon_r t}$ and $e^{\alpha t}$ factors. The annihilation and creation operators are to be written *only* for free lines (which occur for wave function or single particle energy diagrams). The effect of intermediate creation and annihilation operators is taken in account by attaching the factor $(-)^{h+l}$ to the contribution. Time integrations can then be carried out to generate the correct energy denominators. Summation over the indices has to be suitably restricted. All calculations may, perhaps more conveniently, be done by using Hugenholtz's diagrams (section 4).

4. RESOLVENT OPERATOR METHOD: TIME INDEPENDENT THEORY (H1, H2)

The chief aim of this section is to develop general formulae for the energies and wave functions of various types of states of the many body system. For this purpose the time independent method of Hugenholtz is followed. In this method use is made of an expansion of the Green's function. Diagrammatic techniques are further developed and are combined with the powerful methods of complex integration to obtain from the expansion of the Green's function compact expressions for the quantities of interest. These expressions are then used to discuss the properties of the system in some limiting cases which correspond to phenomena like single particle and metastable states, in addition to the ground state properties which are of main interest.

different zero order Hamiltonian to avoid the difficulty in the present case. Validity of this expansion has also been questioned by Cooper (C8) who has shown that under some conditions its radius of convergence is zero.

Operators appropriate to a very large system of Fermions – the so-called Fermi gas – are introduced in the beginning and used throughout this section. However, many of the results are independent of this assumption.

The connection of the present method to the time dependent theory of the previous section is also shown.

Fermi gas: Conventions and definitions

For an infinite system our basic single particle wave functions are plane waves [21]

$$\psi_k = \Omega^{-\frac{1}{2}} \exp(ikx), \tag{4.1}$$

where $\Omega = L^3$ is the large volume in which the large number, viz. N, of the particles are confined. The boundary conditions imply, as usual, that the three components of k can each have values $2\pi n/L$ $(n = 0, \pm 1, \pm 2 \ldots)$. In the present case the interaction operator becomes

$$V = \tfrac{1}{2}\Omega^{-1} \sum_{klmn} \delta_{\mathrm{Kr}}(k + l - m - n) v(k - n) \eta_k^* \eta_l^* \eta_m \eta_n \tag{4.2}$$

where $v(k)$ is the Fourier transform of the central two-body potential

$$v(k) = \int \mathrm{d}^3x\, v(x) \exp(-ikx) \tag{4.3}$$

and the δ_{Kr}-function (Kronecker type) arises from the exponentials in the usual way.

We are interested here in the limit of $N, \Omega \to \infty$ such that $\varrho = N/\Omega$ remains finite. In such a limit the normalization of the operators has to be changed and summations replaced by integrations because the spectrum becomes continuous. We introduce new operators and other quantities.

$$\xi_k = \left(\frac{L}{2\pi}\right)^{\frac{3}{2}} \eta_k; \quad \delta(k - l) = \left(\frac{L}{2\pi}\right)^3 \delta_{\mathrm{Kr}}(k - l) \tag{4.4}$$

so that

$$\{\xi_k, \xi_l\} = \{\xi_k^*, \xi_l^*\} = 0; \; \{\xi_k, \xi_l^*\} = \delta(k - l). \tag{4.5}$$

Also the following replacement is needed

$$\left(\frac{2\pi}{L}\right)^3 \sum_k \to \int_k \equiv \int \mathrm{d}^3k. \tag{4.6}$$

[21] The indices k etc. and coordinate x etc. are actually vectors but no special notation need be used for this. We have chosen the units such that $\hbar = 1$.

In the ground state of the non-interacting system the momentum states are all filled up to a maximum momentum k_F, the Fermi momentum. In momentum space the system occupies a sphere of radius k_F, the Fermi sphere. The particles occupying this sphere are referred to as the Fermi sea.

The Hamiltonian now becomes

$$H = \int_k \frac{|k|^2}{2m} \xi_k^* \xi_k + \tfrac{1}{2}(2\pi)^{-3} \int_{klmn} \delta(k + l - m - n) v(k - n) \xi_k^* \xi_l^* \xi_m \xi_n. \tag{4.7}$$

Corresponding to Eq. (3.23b)

$$V = \tfrac{1}{4} \int_{klmn} v(klmn) \xi_k^* \xi_l^* \xi_m \xi_n \tag{4.8}$$

where

$$v(klmn) = (2\pi)^{-3}[v(k - n) - v(l - m)]\delta(k + l - m - n).$$

The new function $v(klmn)$ has the following properties under the interchange of its index-arguments,

$$v(klmn) = -v(lkmn) = v(mnkl). \tag{4.9}$$

The summation integration is extended over *all* values of momentum both inside and outside the Fermi sphere.

In terms of the new operators the ground state is given by

$$|\Phi_0\rangle = \xi_{k_1}^* \xi_{k_2}^* \ldots \xi_{k_n}^* |0_A\rangle. \tag{4.10}$$

This definition of $|\Phi_0\rangle$ differs from the definition (3.2′) only in normalisation. Following relations hold as $\Omega \to \infty$

$$N = k_F^3 \Omega / 6\pi^2 \tag{4.11}$$

$$E_0 = k_F^5 \Omega / 20\pi^2 M \tag{4.12}$$

where E_0 is the energy of the ground state of the unperturbed system. An arbitrary state of the unperturbed system is written as (cf. sec. 3a):

$$|k_1 k_2 \ldots k_p;\ m_1, m_2 \ldots m_q\rangle = \xi_{k_1}^* \xi_{k_2}^* \ldots \xi_{k_p}^* \xi_{m_1} \ldots \xi_{m_q} |\Phi_0\rangle \tag{4.13}$$

and its conjugate is denoted by $\langle m_q \ldots m_2 m_1;\ k_p \ldots k_2 k_1|$. The convention about the momenta associated with the indices is

$$|k_i| > k_F, \text{ and } |m_i| < k_F.$$

The following table summarises this convention together with the corresponding graphical representation. (The heavy dot, bullet or vertex, represents an interaction.)

Operator	*Represented by*	*Explanation*
ξ_k	$\bullet\!\!\longleftarrow$	annihilates a particle
ξ_m^*	$\bullet\!\!\longrightarrow$	annihilates a hole
ξ_k^*	$\longleftarrow\!\!\bullet$	creates a particle
ξ_m	$\longrightarrow\!\!\bullet$	creates a hole

The resolvent operator [22]

In Chapter I we have shown how various perturbation methods may be looked upon as a search for the poles of the Green's function of the entire system ((I. 2.4) and (I. 2.3)). There a diagonal matrix W_d was defined such that $W = \boldsymbol{M}W_d$ and the level-shift and the wave function were expressed in terms of the $\boldsymbol{M}$-matrix and a suitable limiting process ((I. 2.16) and (I. 2.18)). This was a time-independent method and we return to it now introducing the diagrammatic representation for the matrix element.

Hugenholtz (H1, H2, H3) in conformity with a practice in mathematical theory (S6) and with some other authors, prefers to define a so-called resolvent operator $R(z)$ by means of the equation

$$R(z) = (H - z)^{-1} = (H_0 + V - z)^{-1} = -W(E = z) \text{ (from Ch. I).} \tag{4.14}$$

From now on we shall adopt the name resolvent operator and the definition (14). It can be shown (S6) that this operator exists and is bounded for any non-real value of z.

To show the connection with the time dependent methods of the previous section recall the Schrödinger equation of the perturbed system

$$\mathrm{i}\frac{\mathrm{d}}{\mathrm{d}t}\Psi_{\mathrm{S}}(t) = H\Psi_{\mathrm{S}}(t)$$

$$\Psi_{\mathrm{S}}(t) = U(t)\Psi(0)$$

then

$$\mathrm{i}\frac{\mathrm{d}}{\mathrm{d}t}U = HU. \tag{4.15}$$

The formal solution of which is

$$U(t) = \exp(-\mathrm{i}Ht).$$

[22] The formal theory of the resolvent operator is available in reference (S6). An early application to perturbation theory was made by Kato (K2).

The connection with the $U(t, 0)$ of the last section is given by (3.28) which is rewritten here

$$U(t, 0) = e^{iH_0 t}\, U(t) = e^{-iVt}. \tag{4.16}$$

The matrix $U(t)$ is now connected to the resolvent operator (or Green's function) (14) by means of the relation

$$U(t) = -(2\pi i)^{-1} \oint dz\, R(z) \exp(-izt) \tag{4.17}$$

where the path of integration is a contour described counter clockwise around a sufficiently large portion of the real axis in the complex z-plane as shown in Fig. 15.

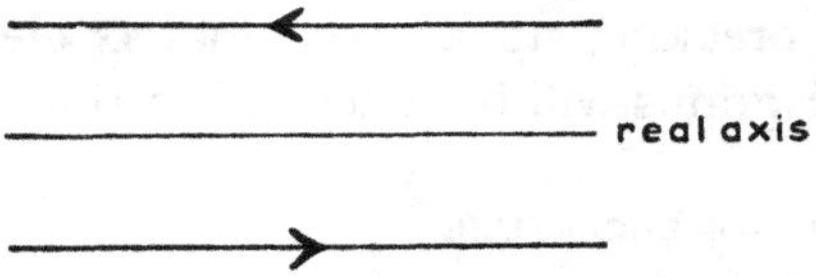

Fig. 15. Contour for integration in the right hand side of (II. 4.17).

To see the truth of (17) consider any function Ψ in Hilbert space which can be expressed as a linear combination of the eigenfunctions Ψ_λ of H belonging to eigenvalue $\mathscr{E}_\lambda$,

$$\Psi = \sum_\lambda a_\lambda \Psi_\lambda$$

then

$$U\Psi = -(2\pi i)^{-1} \oint dz \sum_\lambda \frac{a_\lambda}{\mathscr{E}_\lambda - z} \exp(-izt)\, \Psi_\lambda$$

$$= \sum_\lambda a_\lambda e^{-i\mathscr{E}_\lambda t}\, \Psi_\lambda \equiv e^{-iHt}\Psi \tag{4.18}$$

alternatively one could substitute (17) in the differential equation (15) and show that this together with the boundary condition $U(0) = 1$ are satisfied by the right hand side of (17) (H2).

We shall be interested here in the behaviour of the resolvent operator in a very small neighbourhood of the real axis and in particular in its behaviour close to the singularities.

From (14) we have the identities

$$R(z) = (H_0 - z)^{-1} - (H_0 - z)^{-1} V R(z) \tag{4.19}$$

$$= (H_0 - z)^{-1} - R(z) V (H_0 - z)^{-1}. \tag{4.20}$$

On iterating these we have the series expansion of $R(z)$,

$$R(z) = (H_0 - z)^{-1} - (H_0 - z)^{-1}V(H_0 - z)^{-1} + (H_0 - z)^{-1}V(H_0 - z)^{-1}V(H_0 - z)^{-1} + \dots . \qquad (4.21)$$

The convergence of this series is not at all obvious but has to be investigated for each case separately. Here we shall assume this convergence and turn to a study of the methods of evaluating the matrix elements of $R(z)$ by means of diagrams.

4a. *Matrix elements of the resolvent operator*

The method of breaking up a given matrix element of $R(z)$ into various types of diagrams will be described in this section.

4a (i). Notation and nomenclature

Connected diagrams: A diagram which cannot be divided in two or more parts without cutting any particle or hole line. It should be noted that although all connected diagrams are 'linked' according to the definition of the previous section, *all linked diagrams are not connected*. An example is given in Fig. 23. However, it will be seen later that the 'linked' diagrams occurring in ground state energy expressions happen to be also connected.

Disconnected diagram: A diagram which is not connected according to the above definition. It follows that a disconnected diagram is made up of some *connected parts*.

Component: Same as the *connected part* of a diagram. A disconnected diagram is made up of two or more components.

Ground state diagram: A diagram having no external lines.

Ground state component: A component having no external lines.

Diagonal diagram: A diagram in which the numbers of particle and hole lines before the first interaction and after the last interaction are the same, and in which a product of δ-functions on the momenta of all the particles and holes occurs in the contribution (Fig. (16)).

Vacuum diagram: A diagonal diagram going from vacuum to vacuum, i.e., having no external lines (e.g., Fig. 16a).

Self energy diagram: A diagonal diagram with only one particle or one hole at both ends along with the delta function $\delta^3(k - k')$ or $\delta^3(m - m')$ (e.g., Fig. 16a, b).

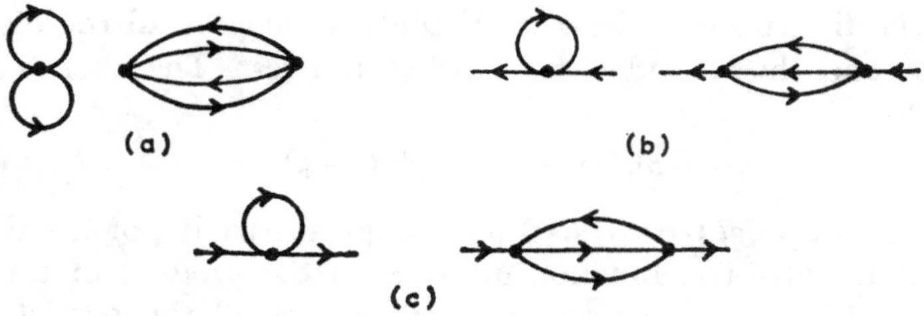

Fig. 16. Examples of basic diagonal diagrams. (a) vacuum diagrams (b) self-energy of a particle and (c) self-energy of a hole.

Diagonal sub-diagram: That part of a diagram, between any two lines drawn perpendicular to the direction of time, which, if taken by itself, would be a diagonal diagram.
Reducible diagram: A diagram which has one or more diagonal sub-diagrams.
Irreducible diagram: A diagram which is not reducible.

The following subscripts applied to an expression mean that only the indicated types of diagrams are to be considered in evaluating that expression:

C	connected
D	completely disconnected
d	diagonal
i	irreducible
id	irreducible diagonal
ind	irreducible non-diagonal
L	linked
nd	non-diagonal.

There are only three basic types of diagonal diagrams: (a) vacuum diagrams, (b) self energy diagrams for a hole, and (c) self energy diagrams for a particle (see Fig. 16). All other diagrams may be expressed in terms of these diagrams by the process of reducing them.

A general matrix element $\langle\beta|R(z)|\alpha\rangle$ of $R(z)$ for the states $|\beta\rangle$,

$|\alpha\rangle$ of the unperturbed system, containing the same number of holes and particles, may be unambiguously written as

$$\langle\beta|R(z)|\alpha\rangle = D_\alpha(z)\delta(\beta - \alpha) + F_{\beta\alpha}(z). \tag{4.22}$$

Where the first term is the sum of contributions of all the diagonal diagrams and the second is the sum of the rest. The operator $D(z)$ defined by

$$\langle\beta|D(z)|\alpha\rangle = D_\alpha\delta(\beta - \alpha) \tag{4.23}$$

is called the *diagonal* part of $R(z)$ (cf., diagonal part W_d of the Green's function (I. 2.6)). The function $\delta(\beta - \alpha)$ is the product of three dimensional delta functions on the momenta of all the particles and holes involved:

$$\delta(\beta - \alpha) = \delta^3(k_1 - k_1') \ldots \delta^3(k_i - k_i')\delta^3(m_1 - m_1') \ldots \delta^3(m_j - m_j').$$

Example of a reducible diagram (Fig. 17)

This diagram has six vertices (6th order) between any two of which we can introduce an intermediate state $|\gamma_i\rangle$. There are five such states. Three of these lead to two delta functions, viz., $\delta(\beta - \gamma_4)$ and

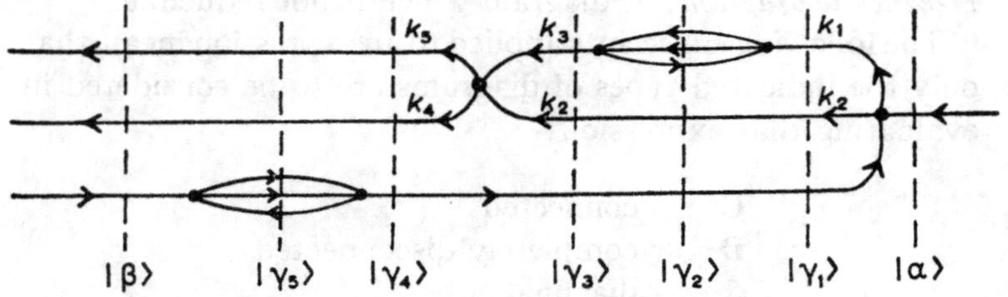

Fig. 17. Example of a non-diagonal reducible diagram.

$\delta(\gamma_3 - \gamma_1)$. Taking out the two indicated parts (diagonal sub-diagrams) and rejoining we have Figure 18a which is an irreducible diagram. Note that between $|\gamma_4\rangle$ and $|\gamma_3\rangle$ the factor is only $\delta(k_2 + k_3 - k_4 - k_5)$, which is not of the form $\delta(\gamma_4 - \gamma_3)$.

4a (ii). EXAMPLE OF CALCULATION OF THE CONTRIBUTION OF A DIAGRAM (Fig. 18)

We consider the second order contributions to $\langle\beta|R(z)|\alpha\rangle$ when

$$|\beta\rangle = |k_2k_3, m\rangle \text{ and } |\alpha\rangle = |k_1\rangle.$$

Taking the second order term from (21) and substituting for V from (8) we have

$$\tfrac{1}{16} \int_{l_1 l_2 l_3 l_4} \int_{n_1 n_2 n_3 n_4} v(l_1 l_2 l_3 l_4) v(n_1 n_2 n_3 n_4)$$
$$\times \langle \Phi_0 | \xi_m^* \xi_{k_3} \xi_{k_2} (H_0 - z)^{-1} \xi_{l_1}^* \xi_{l_2}^* \xi_{l_3} \xi_{l_4} (H_0 - z)^{-1} \ldots$$
$$\xi_{n_1}^* \xi_{n_2}^* \xi_{n_3} \xi_{n_4} (H_0 - z)^{-1} \xi_{k_1}^* | \Phi_0 \rangle. \qquad (4.24)$$

The sum extends over all values of momenta both inside and outside the Fermi sphere, i.e., n_2 and l_2 take on all possible k and m values. The matrix element in the integrand above has been expressed as a ground state to ground state (vacuum to vacuum) transition. All particles and holes created in the *intermediate* states must annihilate each other if the matrix element is to have a non-vanishing contribution. In such a matrix element there must exist as many creation as annihilation operators. These operators furthermore should be paired or associated in the sense that for each creation operator belonging to a hole (or particle) there must be an annihilation operator belonging to the same hole (or particle), which is only to say that the indices on creation and annihilation operators must be paired. Reading always from right to left the creation operator comes first.

Since we have already discussed how diagrams arise it suffices to say that different diagrams result from different possibilities of pairing of various indices and the possibilities of specifying whether

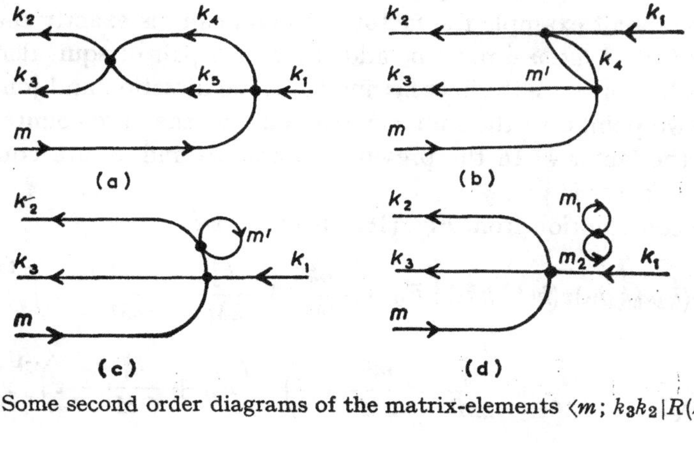

Fig. 18. Some second order diagrams of the matrix-elements $\langle m; k_3 k_2 | R(z) | k_1 \rangle$.

the l and n are inside or outside the Fermi sea. Some of the possible diagrams in the second order are shown in Fig. 18.

Steps involved in calculating the contribution of a given order may now be outlined with the help of this example of the second order matrix element. (a) Write the general expression corresponding to (24). (b) Draw all possible diagrams of this order, as partly done in Fig. 18. Then evaluate the contribution of each diagram separately as follows: (c) Make the pairing of indices appropriate to the diagram. For instance, for the diagram (18a) one has

$$l_1 = k_2,\ l_2 = k_3,\ l_3 = n_2 = k_4,\ l_4 = n_1 = k_5,\ n_3 = m,\ n_4 = k_1.$$

Substituting in (24) we have

$$\tfrac{1}{16} \int_{k_4 k_5} v(k_2k_3k_4k_5)v(k_5k_4mk_1)\langle \Phi_0 | \xi_m^* \xi_{k_3} \xi_{k_2} (H_0 - z)^{-1} \\ \times \xi_{k_5}^* \xi_{k_4}^* \xi_m \xi_{k_1} (H_0 - z)^{-1} \xi_{k_1}^* | \Phi_0 \rangle. \qquad (4.25)$$

(d) Bring the creation and annihilation operators with the same index next to each other. This gives a factor $+1$ or -1 depending on whether the number of permutations needed to accomplish this is even or odd. With this arrangement the expectation value is immediately taken and is equal to one except for the energy denominators. In (25), for diagram Fig. (18a), the factor is $+1$. This is actually the factor $(-)^{h+l}$ of Goldstone (sec. 3b). (e) Important numerical factors are now to be taken into account:

A factor of 4 for each vertex since one obtains identical contributions on interchanging the two ξ's or the two ξ^*'s belonging to the same V. In the present example the factor is 16 and cancels exactly the factor 1/16 in (25). *A factor* $\frac{1}{2}$ must be added for each pair of equivalent lines. Equivalent lines are the lines having the same direction and joining the same two points. In the sum as written above they are counted twice hence the factor $\frac{1}{2}$. In the present example k_4 and k_5 are equivalent and the factor is just $\frac{1}{2}$.

The contribution from Fig. (18a) is therefore

$$\frac{1}{2} \int_{k_4 k_5} v(k_2k_3k_4k_5)v(k_5k_4mk_1) \left(E_0 + \frac{k_2^2}{2M} + \frac{k_3^2}{2M} - \frac{m^2}{2M} - z \right)^{-1} \\ \times \left(E_0 + \frac{k_4^2}{2M} + \frac{k_5^2}{2M} - \frac{m^2}{2M} - z \right)^{-1} \left(E_0 + \frac{k_1^2}{2M} - z \right)^{-1}. \qquad (4.26)$$

Because of the Pauli principle one should require that in (26) the intermediate state $k_4 \neq k_5$. It is clear, however, from the antisymmetry of v that when $k_4 = k_5$ the matrix element itself vanishes. As suggested before this is an example of the general rule which says that one can forget the requirements of the Pauli principle for the intermediate states.

The present example contains all the features needed for calculation of the contribution of any diagram. This example was originally given by Hugenholtz (H1, H2).

4a (iii). $R(z)$ IN TERMS OF DIAGRAMS: THE OPERATOR $G(z)$

It follows from the definitions introduced earlier that any non-diagonal diagram may be constructed in an unique way by inserting appropriate diagonal subdiagrams in an irreducible non-diagonal diagram (**ind**). In terms of the calculational procedure it means that one takes the expression for the appropriate **ind** and in place of the factors $(E_\alpha - z)^{-1}$, $(E_\gamma - z)^{-1}$, $(E_\beta - z)^{-1}$ belonging to the initial, intermediate and final states, substitutes the contributions of appropriate diagonal diagrams without the δ-factors. The sum of the contributions of all non-diagonal diagrams of $\langle\beta|R(z)|\alpha\rangle$ is obtained by taking the sum of the contributions of all irreducible non-diagonal diagrams and substituting for the factors $(E_\alpha - z)^{-1}$, $(E_\gamma - z)^{-1}$ and $(E_\beta - z)^{-1}$ the functions $D_\alpha(z)$, $D_\gamma(z)$ and $D_\beta(z)$ defined in (23). All this is expressed by the formula

$$\{R(z)\}_{\mathbf{nd}} = [-D(z)VD(z) + D(z)VD(z)VD(z) - \ldots]_{\mathbf{ind}}. \tag{4.27}$$

Hence also for $R(z)$ itself,

$$R(z) = D(z) + [-D(z)VD(z) + D(z)VD(z)VD(z) - \ldots]_{\mathbf{ind}}. \tag{4.28}$$

Next it is desirable to express the $D(z)$ operator itself in terms of irreducible diagrams. This requires a prescription for reducing the diagonal diagrams. In the convention of Hugenholtz, the reduced diagonal diagram is to be obtained by removing all the diagonal subdiagrams not containing the first vertex (from the right) of the original diagram. In the reverse process of obtaining a diagonal diagram from the corresponding irreducible diagram one replaces the factors $(E_\beta - z)^{-1}$, $(E_\gamma - z)^{-1}$ by D_β and D_γ of appropriate diagonal subdiagrams, the factor $(E_\alpha - z)^{-1}$ remaining unaltered. Hence the formula

$$D(z) = (H_0 - z)^{-1} + D(z)[-V + VD(z)V - VD(z)V + \ldots]_{\mathrm{id}}(H_0 - z)^{-1}. \quad (4.29)$$

The order of factors in the second term on the right hand side is unimportant since all the factors are diagonal. Therefore the result is independent of the convention defining the process of reduction. This may be looked upon as a formal summation of the series. Some more formal manipulations are needed. Let

$$G(z) = [-V + VD(z)V - VD(z)VD(z)V + \ldots]_{\mathrm{id}}. \quad (4.30)$$

Then (29) becomes

$$D(z) = (H_0 - z)^{-1} + D(z)G(z)(H_0 - z)^{-1}$$

or

$$D(z) = [H_0 - z - G(z)]^{-1}. \quad (4.31)$$

This is one of the basic equations derived by Van Hove (v.H 1) in a slightly different way. One can generate the successive approximations to $D(z)$ and $G(z)$ from (30) and (31) by starting with $G^{(0)}(z) = 0$ in (31).

The Hermitian nature of H implies that

$$R(z^*) = R^*(z), \; D_\alpha(z^*) = D^*_\alpha(z), \; G_\alpha(z^*) = G^*_\alpha(z). \quad (4.32)$$

It has been mentioned earlier that because of the Hermitian nature of H, the resolvent operator $R(z)$ is holomorphic (analytic) for non-real z. From the definition (cf., Eq. (23))

$$\{R(z)\}_{\mathrm{d}} = D(z) \quad (4.33)$$

it follows that $D(z)$ is also holomorphic for non-real z. Equation (31) then implies that $G(z)$ is also holomorphic under the same condition ($\mathrm{Im}\, z \neq 0$). If the spectrum of H_0 is bounded, then from (33) and (31) we conclude further that

$$\lim_{|z|\to\infty} D(z) = 0; \; \lim_{|z|\to\infty} G(z) = 0. \quad (4.34)$$

From the definition (14) of $R(z)$ we have the operator identity

$$R(z') - R(z) = (z' - z)R(z')R(z). \tag{4.35}$$

So that from (33)

$$D(z') - D(z) = (z' - z)\{R(z')R(z)\}_{\mathbf{d}}. \tag{4.36}$$

Using (31) and (28) this gives

$$G(z') - G(z) = (z' - z)\{\{V - VD(z')V + \ldots\}_{\mathbf{ind}} \times D(z')D(z)\{V - VD(z)V + \ldots\}_{\mathbf{ind}}\}_{\mathbf{d}}, \tag{4.37}$$

where $\{V\}_{\mathbf{ind}}$ must be understood to be equal to V. Put $z' = z^*$ in (37). Since $D^*(z) = D(z^*)$ we can conclude that diagonal part of the right hand side is non-negative. Hence it follows that

$$\mathrm{Im}[G_\alpha(z)] \geqslant 0, \text{ for } \mathrm{Im}\, z > 0. \tag{4.38}$$

Also

$$\mathrm{Im}\, D_\alpha(z) > 0. \tag{4.38'}$$

$G_\alpha(z)$ and $D_\alpha(z)$ have singularities only on the real axes where for a finite system they have a large number of poles.

We are here interested only in the limit $\Omega \to \infty$. Thus in (30), the sum over the intermediate states is replaced by an integration and it follows, barring very special conditions, that $G_\alpha(z)$ has no poles but has finite discontinuities for z crossing the real axis in all points of certain intervals which usually depend on α. In most cases these points of discontinuity lie on a portion of real axis from some finite value up to $+\infty$. This enables one to define the real functions $K_\alpha(x)$ and $J_\alpha(x)$ where $z = x + \mathrm{i}\eta$, x and η are real, $\eta > 0$.

$$\lim_{\eta\to 0} G_\alpha(x + \mathrm{i}\eta) = K_\alpha(x) + \mathrm{i}J_\alpha(x) \tag{4.39}$$

$$\lim_{\eta\to 0} G_\alpha(x - \mathrm{i}\eta) = K_\alpha(x) - \mathrm{i}J_\alpha(x). \tag{4.39'}$$

Eq. (38) then implies

$$J_\alpha(x) \geqslant 0. \tag{4.40}$$

The equality sign holds for those values of x (real axis) where $G_\alpha(z)$ is regular. Where $J_\alpha(x) > 0$ both $G_\alpha(z)$ and $D_\alpha(z)$ have finite discontinuities for z crossing the real axis.

4b. *Energies and wave functions of stationary states*

The function $D_\alpha(z)$ may have poles even when $G_\alpha(z)$ does not have any. This happens when the equation $[E_\alpha - z - G_\alpha(z)] = 0$ has a solution. This solution must be real, since $D_\alpha(z)$ has no singularity except on the real axis. We therefore consider the following equation.

$$E_\alpha - x - K_\alpha(x) = 0. \tag{4.41}$$

We shall suppose that this equation has only one root x_α. For x_α to be a pole of $D_\alpha(z)$ it is necessary and sufficient that $J_\alpha(x) = 0$ for x in the neighbourhood of x_α. In this circumstance

$$x_\alpha = E_\alpha - K_\alpha(x_\alpha) \equiv E_\alpha + \Delta E = \mathscr{E}_\alpha$$

which *shows that* x_α *is the perturbed energy of the state obtained from* $|\alpha\rangle$ *by switching on the perturbation* (cf., Ch. I, sec. 2). It is to be shown that this state is a stationary state of the total system. Later on it will be shown that under certain conditions when $J_\alpha(x_\alpha) \neq 0$ the root x_α of (41) corresponds to metastable states of the system.

To obtain the expression for the wave function of a stationary state under the assumption that $D_\alpha(z)$ has a pole at $z = x_\alpha$ one proceeds as follows. Rewrite (28) as

$$\langle\beta|R(z)|\alpha\rangle = \langle\beta|[1 + D(z)\{-V + VD(z)V - \ldots\}_{\text{ind}}]|\alpha\rangle D_\alpha(z). \tag{4.42}$$

The last factor on the right hand side has a pole at x_α while the second factor has a finite discontinuity (barring exceptional circumstances) for z crossing the real axis. One can therefore define two residues of $\langle\beta|R(z)|\alpha\rangle$, one for the upper and one for the lower half plane, thus

$$\mathscr{R}^{\pm}_{x_\alpha}[\langle\beta|R(z)|\alpha\rangle] = \lim_{z\to x_\alpha \pm i0} (z - x_\alpha)\langle\beta|R(z)|\alpha\rangle. \tag{4.43}$$

Explicitly from (42)

$$\mathscr{R}^{\pm}_{x_\alpha}[\langle\beta|R(z)|\alpha\rangle] = -N_\alpha\langle\beta|[1 + D(x_\alpha \pm i0) \times\{-V + VD(x_\alpha \pm i0)V - \ldots\}_{\text{ind}}]|\alpha\rangle,$$

where

$$-N_\alpha = -(1 + G'_\alpha(x_\alpha))^{-1} \tag{4.44}$$

is the residue of $D_\alpha(z)$ in x_α. We define

$$\int_\beta d\beta|\beta\rangle\, \mathscr{R}^{\pm}_{x_\alpha}\langle\beta|R(z)|\alpha\rangle = \mathscr{R}^{\pm}_{x_\alpha}[R(z)\,|\alpha\rangle]. \tag{4.45}$$

From (35) we have

$$R(z)\,|\alpha\rangle - R(z')\,|\alpha\rangle = (z - z')R(z)R(z')\,|\alpha\rangle.$$

As a function of z' both sides have a pole of the type described above. Equating the residues in $z' = x_\alpha$ on both sides we have

$$R(z)\{\mathscr{R}^{\pm}_{x_\alpha}[R(z')\,|\alpha\rangle]\} = \frac{1}{x_\alpha - z}\{\mathscr{R}^{\pm}_{x_\alpha}[R(z')\,|\alpha\rangle\}. \tag{4.46}$$

Or from the relation (17) between $U(t)$ and $R(z)$, we have

$$U(t)\{\mathscr{R}^{\pm}_{x_\alpha}[R(z')\,|\alpha\rangle]\} = \exp(-\mathrm{i}x_\alpha t)\{\mathscr{R}^{\pm}_{x_\alpha}[R(z')\,|\alpha\rangle]\}. \tag{4.46'}$$

The equations (46) show that the state defined by (45) is the stationary eigenstate with energy $x_\alpha = E_\alpha - K_\alpha(x_\alpha)$.

It can be further shown (v.H1) that the normalization constant is $N_\alpha^{-\frac{1}{2}}$. Hence the incoming (plus) and outgoing (minus) stationary states are given by

$$\begin{aligned} |\Psi_\alpha\rangle^{\pm} &= -N_\alpha^{-\frac{1}{2}}\mathscr{R}^{\pm}_{x_\alpha}[R(z)\,|\alpha\rangle] \\ &= +N_\alpha^{+\frac{1}{2}}[1 + D(x_\alpha \pm \mathrm{i}0)\{-V + VD(x_\alpha \pm \mathrm{i}0)V - \ldots\}_{\mathrm{ind}}]\,|\alpha\rangle. \end{aligned} \tag{4.47}$$

It has further been proved (v.H1) that if for all α, $J_\alpha(x_\alpha) = 0$, then the states $\Psi^{(+)}_\alpha$ form a complete orthonormal set and so do $\Psi^{(-)}_\alpha$.

4c. *Convolutions and decomposition of diagrams*

First we discuss some mathematical properties of the convolution integrals which will be of much use in the subsequent development.

Let $f(z)$ and $g(z)$ be two functions holomorphic for non-real z, such that they behave like $1/z$ for $|z| \to \infty$ (cf. (34)). Then a new function, $f(z) \ast g(z)$, can be defined by means of the integral

$$f(z) \ast g(z) = -(2\pi\mathrm{i})^{-1} \oint \mathrm{d}\zeta\, f(z - \zeta)g(\zeta) \tag{4.48}$$

where the path of integration is shown below. It is drawn so as to enclose the singularities of the integrand on the real ζ-axis but not enclosing the singularities on the line through z parallel to the real axis.

This asymmetry is only apparent, since the property that $zf(z)$ and $zg(z)$ are

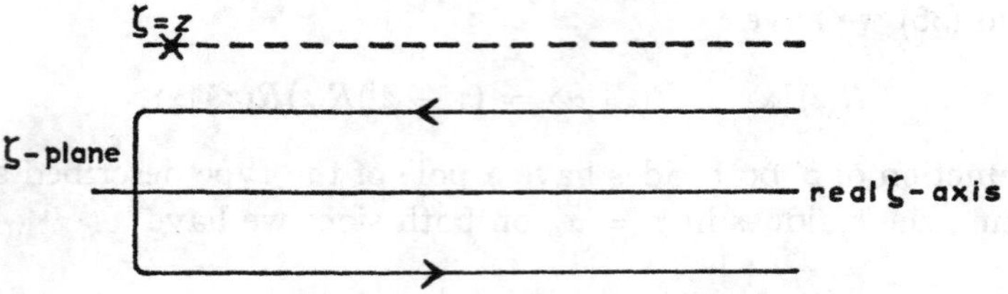

Fig. 19. Contour defining the convolution in (II. 4.48).

bounded for large $|z|$ the contour can be deformed so as to encircle the line through z parallel to real axis. Hence

$$f(z) * g(z) = g(z) * f(z). \tag{4.49}$$

Consider now the quantity F defined as follows:

$$F(a_n \dots a_0; b_m \dots b_0) = \prod_{k=0}^{n} (a_k - z)^{-1} * \prod_{l=0}^{m} (b_l - z)^{-1} \tag{4.50}$$

where a_k and b_l are real numbers. Using (48) we can connect F of (50) to quantities involving a lesser number of constants. Thus

$$(a_n + b_m - z)^{-1}[F(a_{n-1} \dots a_0; b_m \dots b_0) + F(a_n \dots a_0; b_{m-1} \dots b_0)] = F(a_n \dots a_0; b_m \dots b_0), \tag{4.51}$$

for the left hand side is

$$= -(2\pi i)^{-1} \oint d\zeta\, (a_n + b_m - z)^{-1} \left(\frac{1}{b_m - \zeta} + \frac{1}{a_n - z + \zeta} \right) \frac{1}{a_{n-1} - z + \zeta} \cdots \frac{1}{a_0 - z + \zeta} \frac{1}{b_{m-1} - \zeta} \cdots \frac{1}{b_0 - \zeta}.$$

As functions of z the contributions of diagrams have the form of products of the above type. The theorem stated above can be used to express the contribution of a set of disconnected diagrams as a convolution of the contributions of their connected parts in the manner to be described below.

As before a state $|\alpha\rangle$ of the basic system is obtained by applying an appropriate number of creation and annihilation operators on the 'vacuum' state $|\Phi_0\rangle$. A state $|\alpha\beta\rangle$ is said to arise when one applies the same operators in the same order to the state $|\beta\rangle$. Or if α_{op} stands for a string of creation and annihilation operators such that $\alpha_{op}|\Phi_0\rangle = |\alpha\rangle$, then $\alpha_{op}|\beta\rangle = |\alpha\beta\rangle$. The conjugate of $|\alpha\beta\rangle$ is $\langle\beta\alpha|$. The states $|\alpha'\rangle$,

$|\beta'\rangle$ and the state $|\alpha'\beta'\rangle$ are supposed to be related in the same way. The composite state contains all the particles and holes of the constituent states. It can, of course, be defined only if none of the particles or holes in the two states have the same momentum (occupy the same single particle state).

Let $\langle\alpha'|\mathscr{A}(z)|\alpha\rangle$ represent the contribution of a diagram $\mathscr{A}$ of order n to the matrix element $\langle\alpha'|R(z)|\alpha\rangle$ and $\langle\beta'|\mathscr{B}(z)|\beta\rangle$ the contribution of a diagram $\mathscr{B}$ of order m to the matrix element $\langle\beta'|R(z)|\beta\rangle$. The two diagrams $\mathscr{A}$ and $\mathscr{B}$ can be placed side by side and looked upon as an unconnected single diagram of order $n+m$. There are $(n+m)!/n!m!$ distinct diagrams which can be formed in this way corresponding to the number of distinct (time) orderings of the n and m indices with reference to each other. Each time ordering gives rise to different energy denominators. The important result which we now wish to prove is that if $\langle\beta'\alpha'|\mathscr{C}(z)|\alpha\beta\rangle$ denotes the sum of contributions from all composite diagrams formed in this way, then

$$\langle\beta'\alpha'|\mathscr{C}(E_0+z)|\alpha\beta\rangle = \langle\alpha'|\mathscr{A}(E_0+z)|\alpha\rangle * \langle\beta'|\mathscr{B}(E_0+z)|\beta\rangle. \tag{4.52}$$

To prove this relation we observe that the only part of the contributions of $\mathscr{A}$ and $\mathscr{B}$ affected by the convolution is the product of energy denominators,

$$\mathscr{A}: \prod_{k=0}^{n} (a_k - z)^{-1}; \quad \mathscr{B}: \prod_{l=0}^{m} (b_l - z)^{-1},$$

where a_k and b_l are the energies of the intermediate states relative to E_0. Contributions to $\mathscr{C}$ are in form of ordered products of $n+m+1$ factors

$$(a_k + b_l - z)^{-1}; \; k = 0 \ldots n, \; l = 0 \ldots m,$$

with the rule that if $(a_k + b_l - z)^{-1}$ and $(a_{k'} + b_{l'} - z)^{-1}$ are consecutive factors, the first one to the left, then either $k = k'$, $l = l' + 1$ or $k = k' + 1$, $l = l'$. The ordering is thus uniquely defined and there are $(n+m)!/n!m!$ terms. We denote the sum of these products by $F'(a_n \ldots a_1 a_0; b_m \ldots b_1 b_0)$. To connect it to the products of lower order we observe that in each product the extreme left member is always $(a_n + b_m - z)^{-1}$. Thereafter the second factor can be either $(a_n + b_{m-1} - z)^{-1}$ or $(a_{n-1} + b_m - z)^{-1}$, according to the rule; both cases occur. The original ordered product can therefore be written as the factor $(a_m + b_m - z)^{-1}$ times ordered products of lower order whose first terms are $(a_n + b_{m-1} - z)^{-1}$ and $(a_{n-1} + b_m - z)^{-1}$. That is,

$$F'(a_n \ldots a_0; b_m \ldots b_0) = (a_n + b_m - z)^{-1}[F'(a_{n-1} \ldots a_0; b_m \ldots b_0) + F'(a_n \ldots a_0; b_{m-1} \ldots b_0)]$$

which is exactly the relation (51) satisfied by the convolution F defined by (50). Furthermore from the definitions

$$F(a_0; b_l \dots b_0) = F'(a_0; b_l \dots b_0)$$
$$F(a_k \dots a_0; b_0) = F'(a_k \dots a_0; b_0).$$

Hence

$$F = F'$$

and one concludes that (52) is true.

In this derivation, it should be emphasised, essential use has been made of the fact that the Pauli principle [23] for the intermediate states can be ignored. This enables us to disregard all intermediate creation and annihilation operators. The other important thing was the measurement of the energy of the intermediate states with reference to the energy of the unperturbed ground state $|\Phi_0\rangle$.

EXAMPLE

Three ways of combining the diagrams $\mathscr{A}$ and $\mathscr{B}$ of first and second order respectively are shown in Fig. 20. In this example the above result may be verified by explicit calculation.

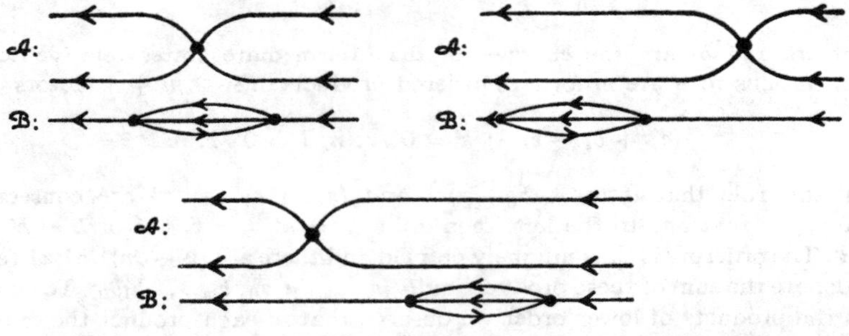

Fig. 20. Three ways of combining a first and a second order diagram. Contribution of all the three composite diagrams may be represented by a convolution over the contribution of $\mathscr{A}$ and $\mathscr{B}$. See text.

[23] Reference here is to that effect of Pauli principle which forbids repetition in the intermediate states of those states which are already occupied (in the chosen configuration). It does not mean that anticommutation properties of the operators are neglected altogether.

4d. *Volume dependence of various quantities*

Consider contributions to the matrix element $\langle\beta|R(z)|\alpha\rangle$ from various diagrams. Each energy denominator has a term proportional to Ω in it. $E_\alpha = E_0 + E'_\alpha$, where E'_α is the energy of the holes and particles making up the state $|\alpha\rangle$, it follows that the contributions to $\langle\beta|R(E_0 + z)|\alpha\rangle$ will have no Ω dependence in the energy denominators and will be easier to discuss. Apart from this $\langle\beta|R(E_0 + z)|\alpha\rangle$ and $\langle\beta|R(z)|\alpha\rangle$ contain the same information.

Consider now various types of contributions to $\langle\beta|R(E_0 + z)|\alpha\rangle$ (refer to section 4a).

(a) Connected diagrams with external lines: A delta function $\delta(\alpha - \beta)$ occurs. If the summations are replaced by integration we get a quantity independent of Ω in the limit $\Omega \to \infty$ (corrections are $O(1/\Omega)$).

(b) Connected ground state diagrams: Since there are no external lines one has an extra δ-function, which is dependent on the other delta function through the conservation of momentum at each vertex. This leads to a factor $\delta(0)$ which gives a factor $\Omega/8\pi^3$. The other integrations are as before and hence the leading term here is proportional to Ω.

(c) Diagrams containing one or more ground state components: If n is the number of such components the contribution is proportional to Ω^n.

Hence the expansion of $\langle\beta|R(E_0 + z)|\alpha\rangle$ contains arbitrarily high powers of Ω and is on the face of it useless. However, on performing sums over unlinked diagrams these Ω^n-dependent ($n > 1$) parts were seen to cancel, thus removing one important undesirable feature of the expansion.

Recalling the work of the previous section, we find a way of separating Ω-independent and Ω-dependent contributions to the general matrix element $\langle\alpha'|R(z)|\alpha\rangle$ for $R(z)$ when $\alpha' \neq \alpha$. Let us define by $\langle\alpha'|\bar{R}(z)|\alpha\rangle$ the sum of contributions of all those diagrams of $\langle\alpha'|R(z)|\alpha\rangle$ which do not have any ground state component, that is, those independent of Ω. Now any arbitrary diagram belonging to $\langle\alpha'|R(z)|\alpha\rangle$ can be constructed from a suitable diagram of the above type by placing along with it, in a suitable way, a certain desired number of ground state components. Hence by the work of the previous section we have

$$\langle\alpha'|R(E_0+z)|\alpha\rangle = \langle\alpha'|\bar{R}(E_0+z)|\alpha\rangle * D_0(E_0+z). \qquad (4.53)$$

The left hand side represents the total contribution from all possible diagrams. The first term on the right hand side is the total contribution from all diagrams without ground state components. The second term on the right hand side is the sum of contributions of all ground state diagrams. The function D_0 contains all the Ω-dependent contributions, which can be of arbitrarily high powers of Ω. This separation of Ω-dependence is the great merit of equation (53).

4e. *Discussion of energy expressions*

4e (i). Ground state energy: quantities D_0 and $\bar{G}_0$

The Integral Equation for $D_0(z)$

The poles of $D_0(z)$ correspond to the eigenvalues of the energy of the perturbed system. As it stands $D_0(z)$ contains contributions also from unconnected diagrams. The next step in our study is to express $D_0(z)$ in terms only of connected diagrams and to study its poles. For this purpose we derive an integral equation for $D_0(z)$.

Taking the diagonal elements with reference to $|\Phi_0\rangle$ of both sides in (20) and replacing z by E_0+z, we have on introducing the unit matrix between $R(z)$ and V

$$\begin{aligned} D_0(E_0+z) &= -\frac{1}{z} + \int_\alpha \langle\Phi_0|R(E_0+z)|\alpha\rangle\,\langle\alpha|V|\Phi_0\rangle\frac{1}{z} \\ &= -z^{-1}[1 - D_0(E_0+z) * \{z^{-1}\langle\Phi_0|V|\Phi_0\rangle \\ &\qquad - \textstyle\int_\alpha'\langle\Phi_0|\bar{R}(E_0+z)|\alpha\rangle\,\langle\alpha|V|\Phi_0\rangle\}] \end{aligned}$$

where $\int_\alpha'$ indicates that in the summation (integration) $\alpha \neq \Phi_0$. Finally

$$\begin{aligned} zD_0(E_0+z) &= -1 + D_0(E_0+z) * z^{-1}\bar{G}_0(E_0+z) \\ &= -1 - (2\pi \mathrm{i})^{-1}\oint \mathrm{d}\zeta\,\zeta^{-1}\bar{G}_0(E_0+\zeta)D_0(E_0+z-\zeta) \qquad (4.54) \end{aligned}$$

so that

$$\bar{G}_0(E_0+z) = -\langle\Phi_0|V|\Phi_0\rangle + z^{-1}\textstyle\int_\alpha'\langle\Phi_0|\bar{R}(E_0+z)|\alpha\rangle\,\langle\alpha|V|\Phi_0\rangle$$

$$\bar{G}_0(z) = -\langle\Phi_0|V|\Phi_0\rangle - \frac{1}{E_0-z}\int_\alpha'\langle\Phi_0|\bar{R}(z)|\alpha\rangle\,\langle\alpha|V|\Phi_0\rangle$$

or

$$\bar{G}_0(z) = \langle \Phi_0 | [-V + V(H_0 - z)^{-1}V - \ldots]_C | \Phi_0 \rangle, \tag{4.55}$$

where the subscript **C** means that only the connected diagrams are taken into account. The connected diagrams in general contain ground state sub-diagrams. With methods previously discussed we can express $\bar{G}_0(z)$ in terms of irreducible diagrams only. Starting from irreducible ground state diagrams one can construct in an unambiguous way all diagrams of (55) by inserting suitable diagonal sub-diagrams between any two successive points. These diagonal sub-diagrams are not ground state diagrams. Therefore, we define for each $\alpha \neq \Phi_0$ the sum $\bar{D}_\alpha(z)$ of the contribution to $D_\alpha(z)$ which do not contain ground state components. Then

$$\bar{G}_0(z) = \langle \Phi_0 | [-V + V\bar{D}(z)V - V\bar{D}(z)V\bar{D}(z)V + \ldots]_{\mathrm{idC}} | \Phi_0 \rangle. \tag{4.56}$$

The integral equation (54) for $D_0(z)$ is now to be solved. We shall do so again in the limit $\Omega \to \infty$.

Properties of $\bar{G}_0(E_0 + z)$

In the limit $\Omega \to \infty$ $\bar{G}_0(E_0 + z)$ is proportional to Ω. As discussed before $\bar{G}_0(E_0 + z)$ has no poles. Its only singularities are on a cut on the real axis from a point B to $+\infty$. In each point of this cut the function has a finite discontinuity for z-crossing the real axis. The function can be continued analytically across the real axis from above or below the real axis. Then B is a branch point of the function. B is assumed to be on the positive real axis or the origin.

The properties enumerated above have been *assumed* for $\bar{G}_0$. In each particular case it is to be shown that they actually hold. Nevertheless it seems that these are quite general conditions to impose on $\bar{G}_0$ in order to get physically meaningful results. For instance if $\bar{G}_0$ has singularities on negative real axis then D_0 will have to have singularities extending to $-\infty$ if (54) has a solution. This would mean that there was no lower bound to the perturbed energy – a result inadmissible on physical grounds.

One defines

$$J_0(x) = -\tfrac{1}{2}\mathrm{i} \lim_{\eta \to 0} [\bar{G}_0(x + \mathrm{i}\eta) - \bar{G}_0(x - \mathrm{i}\eta)]; \ \eta > 0 \tag{4.57}$$

and if B is at the origin an assumption has to be made that

$$|\bar{J}_0(E_0 + x)| < Cx^{\varepsilon+1} \text{ for } x > 0 \tag{4.58}$$

where C and ε are positive constants.

Writing further

$$\bar{G}_0(E_0 + z) = - \langle \Phi_0 | V | \Phi_0 \rangle + g(z)$$

we have in virtue of (56) that $zg(z)$ is bounded for $|z| \to \infty$ (cf. (34)). This allows us to deform the contour C in the Cauchy formula

$$g(z) = (2\pi i)^{-1} \int_C d\zeta\, g(\zeta)(\zeta - z)^{-1} \quad \text{for Im } z \neq 0.$$

into a contour around the singular points of $g(z)$ on the positive real axis (Fig. 21).

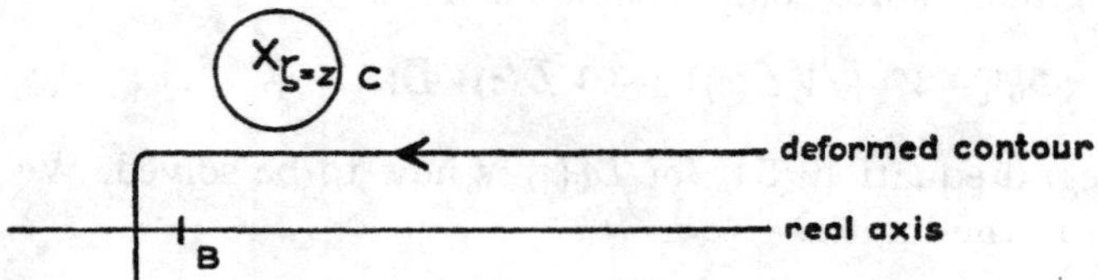

Fig. 21. Deformed contour for integral involving $g(z)$. See discussion below (II. 4.58).

This with (57) gives

$$\bar{G}_0(E_0 + z) = \pi^{-1} \int_0^\infty d\xi\, \bar{J}_0(E_0 + \xi)(\xi - z)^{-1} - \langle \Phi_0 | V | \Phi_0 \rangle. \tag{4.59}$$

This is a type of dispersion relation connecting the function $\bar{G}_0$ to an integral over its imaginary part. Taking derivatives on both sides of (59)

$$\bar{G}_0'(E_0 + z) = \pi^{-1} \int_0^\infty d\xi\, \bar{J}_0(E_0 + \xi)(\xi - z)^{-2}. \tag{4.60}$$

Thus both $\bar{G}_0$ and its derivative exist at the origin. To put the integral equation (54) in a simpler form, we define a function

$$h(z) = (\bar{G}_0(E_0 + z) - \bar{G}_0(E_0))z^{-1}, \tag{4.61}$$

which again has no singularities except for a cut on the real axis from B to $+\infty$, then,

$$h(0) = G_0'(E_0).$$

Substituting (61) in (54) and replacing z by $z - \bar{G}_0(E_0)$ with the definition

$$f(z) = D_0(E_0 - \bar{G}_0(E_0) + z) \tag{4.62}$$

we finally have the required integral equation which is

$$\begin{aligned} zf(z) &= -1 - (2\pi i)^{-1} \oint d\zeta\, h(\zeta)f(z-\zeta) \\ &= -1 - h(z) * f(z). \end{aligned} \tag{4.63}$$

Since $zh(z)$ and the required solution $zf(z)$ are both bounded for $|z| \to \infty$, it follows that

$$\begin{aligned} zf(z) &= -1 - f(z) * h(z) \\ &= -1 - (2\pi i)^{-1} \oint d\zeta\, f(\zeta)h(z-\zeta). \end{aligned} \tag{4.64}$$

Since for $|z| \to \infty$ the second term on the right hand side goes to zero, we have

$$-(2\pi i)^{-1} \oint dz\, f(z) = 1. \tag{4.65}$$

(This can be derived also from (64).)

From the physical point of view we are interested only in those solutions of (63) which are holomorphic outside the real axis and bounded for large $|z|$. It has been shown by Hugenholtz, to whose work (H1) we refer for details of proof, that (63) has a *unique* solution of this type which may be written as

$$f(z) = -\frac{N_0}{z} + \psi(z) \tag{4.66}$$

where

$$N_0 = \exp(-G_0'(E_0)) \tag{4.67}$$

and $\psi(z)$ is a function which has no other singularities except for a cut on the real axis from B to $+\infty$ (cf., $h(z)$). In terms of the discontinuity $\varphi(x)$ of $\psi(z)$

$$2\pi i\varphi(x) = \lim_{\eta\to 0} [\psi(x+i\eta) - \psi(x-i\eta)] \tag{4.68}$$

the Cauchy theorem gives us

$$\psi(z) = \int_0^\infty d\xi\, \varphi(\xi)(\xi - z)^{-1}. \tag{4.69}$$

The solution of the original integral equation (54) now becomes

$$D_0(z) = \frac{\exp(-\bar{G}_0'(E_0))}{E_0 - \bar{G}_0(E_0) - z} + \int_0^\infty \frac{\varphi(\xi)\, d\xi}{\xi - z + E_0 - \bar{G}_0(E_0)} \tag{4.70}$$

where $\varphi(\xi)$ is put explicitly in terms of J_0 and N_0.

The quantities $\bar{G}_0'(E_0)$ and $J_0(E_0 + z)$ are proportional to Ω for large Ω. The merit of (70) is to have expressed $D_0(z)$ in terms of such quantities alone.

We are interested in the behaviour near the poles, which occur at

$$x_0 \equiv E_0 + \Delta E_0 = E_0 - \bar{G}_0(E_0). \tag{4.71}$$

We see that the perturbed energy is proportional to Ω for large Ω. This result can be anticipated on a physical argument which is as follows: the system is divided into large enough cells so that the interactions across the boundaries may be neglected. Then the energy of each cell would simply add up, giving a total energy proportional to Ω. Here the result is rigorously established, the two forms of ΔE_0 are obtained from (55) and (56). Equation (55) is the well-known linked cluster formula.

The present method is more general in that it allows one to discuss the Ω dependence of quantities other than the ground state energy shift.

It was remarked earlier that $D_0(z)$ involves arbitrarily high powers of Ω. This fact is reflected in the occurrence of the factor N_0 in $D_0(z)$ which depends exponentially on Ω. (The residue at the pole of $D_0(z)$ is $-N_0$.) $N_0^{\frac{1}{2}}$ is the normalization constant for the wave function $|\Psi_0\rangle$ of the perturbed ground state. Hence N_0 is the probability of finding the ground state $|\Phi_0\rangle$ in the actual wave function (47). One can understand this result again on the basis of the physical argument used for explaining the Ω dependence of the energy. It is clear that the total wave function is a product of the wave functions of the cells and must therefore depend exponentially on Ω.

4e (ii). EXCITED STATE ENERGIES

The derivation of these quantities is entirely analogous to the foregoing except that the expectation values are now taken with respect to $|\alpha\rangle \neq |\Phi_0\rangle$. The starting point is again (53) from where we have

$$D_\alpha(E_0 + z) = \bar{D}_\alpha(E_0 + z) * D_0(E_0 + z). \tag{4.72}$$

Instead of the integral equation for D_0 we are now interested more in the expression for $\bar{D}_\alpha$ which, in analogy with (31), is given by

$$\bar{D}_\alpha(E_0 + z) = (E'_\alpha - z - \bar{G}_\alpha(E_0 + z))^{-1} \tag{4.73}$$

where $E'_\alpha = E_\alpha - E_0$ and

$$\bar{G}_\alpha(E_0 + z) = \langle\alpha|[-V + V\bar{D}(E_0 + z)V - \ldots]_{\text{idL}}|\alpha\rangle \tag{4.74}$$

where **idL** means that only irreducible diagonal diagrams without ground state components contribute. The subscript **L** is to remind ourselves that these are the so-called 'linked clusters'.

The important difference now is that both $\bar{D}_\alpha$ and $\bar{G}_\alpha$ are defined by means of diagrams without ground state components and are independent of Ω for $\Omega \to \infty$.

From (74) as before we can conclude that $\bar{G}_\alpha(E_0 + z)$ has no poles barring exceptional circumstances, as $\Omega \to \infty$. There will, however, be one or more cuts along the real axis. We further have, as in (32)

$$\bar{G}_\alpha(z^*) = \bar{G}^*_\alpha(z)\,;\, \bar{D}_\alpha(z^*) = \bar{D}^*_\alpha(z)$$

and can define the real functions $\bar{K}_\alpha$ and $\bar{J}_\alpha$ as in (39)

$$\lim_{\eta\to 0} \bar{G}_\alpha(x + i\eta) = \bar{K}_\alpha(x) + i\bar{J}_\alpha(x),\ \eta > 0$$

$$\lim_{\eta\to 0} \bar{G}_\alpha(x - i\eta) = \bar{K}_\alpha(x) - i\bar{J}_\alpha(x),\ \eta < 0.$$

The singular points of $\bar{G}_\alpha(E_0 + z)$ are when $\bar{J}_\alpha(E_0 + x) \neq 0$. These are also the singular points of $\bar{D}_\alpha(E_0 + z)$ from (73). In addition $\bar{D}_\alpha(E_0 + z)$ has a pole if

$$E'_\alpha - x - \bar{K}_\alpha(E_0 + x) = 0 \tag{4.75}$$

has a root in the neighbourhood of which $\bar{J}_\alpha(E_0 + x) = 0$. This is entirely analogous to (41) except for the fact of different dependence on Ω which is an important advantage.

Equation (75) has at least one root. For simplicity we investigate the case when there is only one root $\bar{E}_\alpha$ of this equation. The condition $\bar{J}_\alpha(E_0 + x) = 0$ is, however, too strong. For our purposes it is sufficient that for small x

$$\bar{J}_\alpha(E_0 + \bar{E}_\alpha + x) = O(|x|^{1+\varepsilon}),\quad \varepsilon > 0. \tag{4.76}$$

When this is the case we would still continue to call $\bar{E}_\alpha$ a pole of $\bar{D}_\alpha(E_0 + x)$ even though strictly speaking it is not a true pole.

The condition (76) ensures that both $\bar{G}_\alpha(E_0 + z)$ and its derivative $\bar{G}'_\alpha(E_0 + z)$ exist at $\bar{E}_\alpha$ and are finite. From (75) and (76)

$$\bar{E}_\alpha = E'_\alpha - \bar{G}_\alpha(E_0 + \bar{E}_\alpha). \tag{4.77}$$

The residue of $\bar{D}_\alpha(E_0 + z)$ at $\bar{E}_\alpha$, $-\bar{N}_\alpha$, is given by

$$\bar{N}_\alpha^{-1} = 1 + \bar{G}'_\alpha(E_0 + \bar{E}_\alpha). \tag{4.78}$$

Equation (72) allows us to express the pole $x_\alpha - E_\alpha$ and the residue $-N_\alpha$ of $D_\alpha(E_\alpha + z)$ in terms of the corresponding quantities for $\bar{D}_\alpha$ and D_0, leading to the equations

$$N_\alpha = N_0 \bar{N}_\alpha \tag{4.79}$$

$$x_\alpha = x_0 + \bar{E}_\alpha. \tag{4.80}$$

(It can be proved that (76) implies that the discontinuity in $\bar{D}_\alpha(E_0 + z)$ at a point x on the real axis in the neighbourhood of x_α behaves like $O(|x - \bar{E}_\alpha|^{\varepsilon-1})$ and hence the same holds for the discontinuity of $D_\alpha(z)$ in the neighbourhood of x_α. The use of the word pole must be understood in the same sense.)

The equations (38′) for $D_\alpha(z)$ and $D_0(z)$ imply that both N_α and N_0 are positive, hence $\bar{N}_\alpha$ is also positive.

Comparison of (79) and (80) with discussions associated with (46) shows that since x_α is the energy of the perturbed system in the state $|\Psi_\alpha\rangle^\pm$, *$\bar{E}_\alpha$ is the energy of the state $|\Psi_\alpha\rangle^\pm$ compared to the energy x_0 of the perturbed ground state $|\Psi_0\rangle$. That is, $\bar{E}_\alpha$ is the true or observable excitation energy of the state $|\Psi_\alpha\rangle^\pm$.* In field theory where energy x_0 of the physical vacuum is unobservable $\bar{E}_\alpha$ is the only directly measurable (hence physically meaningful) quantity. It is noteworthy that it is independent of Ω or the size of the system.

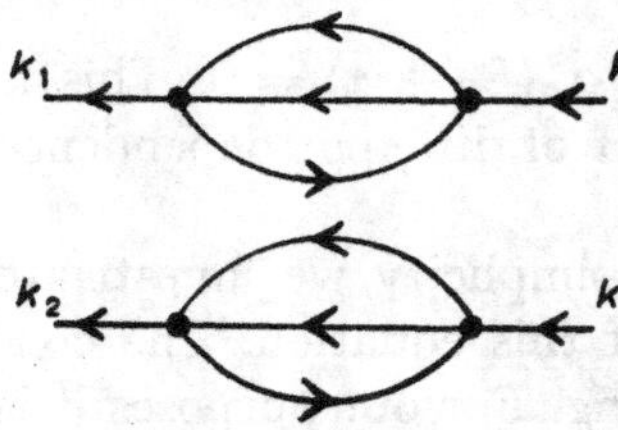

Fig. 22. Diagonal diagrams of a two particle state.

When the state $|\alpha\rangle$ which goes into the physical state $|\Psi_\alpha\rangle^\pm$ contains several particles and holes, the diagonal diagrams contributing to $\bar{D}_\alpha(E_0+z)$ are not connected, they are always composed of one particle state or one hole state. As an example if $|\alpha\rangle = |k_1k_2\rangle$, $|\gamma_1\rangle = |k_1\rangle$, $|\gamma_2\rangle = |k_2\rangle$ then the diagonal diagrams are of the type shown in Fig. 22. So that by an application of the convolution theorem one has

$$\bar{D}_\alpha(E_0 + z) = \bar{D}_{\gamma_1}(E_0 + z) * \bar{D}_{\gamma_2}(E_0 + z) \tag{4.81}$$

and with reasoning similar to that used in deriving (79) and (80) we have

$$\bar{N}_\alpha = \bar{N}_{\gamma_1}\bar{N}_{\gamma_2}; \ \bar{E}_\alpha = \bar{E}_{k_1} + \bar{E}_{k_2}. \tag{4.82}$$

The last equation expresses the additivity of the perturbed excitation energies of two particles. The argument clearly extends to an arbitrary number of particles and holes.

4f. *Discussion of wave function expressions*

The expression for a perturbed state $|\Psi_\alpha\rangle^\pm$ was derived in terms of the residues of a certain quantity involving the resolvent operator and the wave function $|\alpha\rangle$ of the unperturbed system. Now we wish to find some operators $O_\alpha^\pm$ such that

$$|\Psi_\alpha\rangle^\pm = O_\alpha^\pm \, |\Phi_0\rangle \tag{4.83}$$

(cf., (I. 2.18) and connected remarks).

In particular we want the operator that converts the ground state of the basic system to the true ground state.

$$|\Psi_0\rangle^+ = |\Psi_0\rangle^- = |\Psi_0\rangle = O_0 |\Phi_0\rangle. \tag{4.84}$$

It is also of interest to find the operators $\hat{O}_\alpha^\pm$ which "create" the state $|\Psi_\alpha\rangle^\pm$ from the perturbed ground state $|\Psi_0\rangle$

$$|\Psi_\alpha\rangle^\pm = \hat{O}_\alpha^\pm \, |\Psi_0\rangle. \tag{4.85}$$

This section is devoted to constructing explicit forms for these operators in terms of diagrams.

We define an operator $A(z)$ by the equation

$$A_\alpha(z) = \int_\beta \langle\beta|R(z)|\alpha\rangle\beta_{\text{op}}, \tag{4.86}$$

where β_{op} is defined by $\beta_{\text{op}}|\Phi_0\rangle = |\beta\rangle$. Then it follows that

$$A_\alpha(z)\,|\Phi_0\rangle = R(z)\,|\alpha\rangle. \tag{4.87}$$

$A_\alpha(z)$ contains a factor $D_\alpha(z)$ which has a pole at $z = x_\alpha$. Two residues for operator $A_\alpha(z)$ can be defined through those of the individual matrix elements on the right hand side. Then the operators of (83) are obviously, from (47),

$$O_\alpha^\pm = -N_\alpha^{-\frac{1}{2}}\mathscr{R}_{x_\alpha}^\pm[A_\alpha(z)]. \tag{4.88}$$

Now we wish to obtain expressions arising from here which shall allow us to conclude more about the nature of these operators.

4f (i). GROUND STATE WAVE FUNCTION

Consider the operator

$$A_0(z) = \int_\beta \langle\beta|R(z)|\Phi_0\rangle \beta_{\text{op}} = D_0(z) + \int_\beta' \langle\beta|R(z)|\Phi_0\rangle \beta_{\text{op}} \quad (4.89)$$

so that by (42) we have

$$\langle\beta|R(z)|\Phi_0\rangle$$
$$= \langle\beta|[1 + \{-D(z)V + D(z)VD(z)V - \ldots\}_{\text{ind}}]|\Phi_0\rangle D_0(z). \quad (4.90)$$

From (30) we can obtain $G_0(z) = \langle\Phi_0|G(z)|\Phi_0\rangle$ which is similar to the first factor on the right hand side. In particular the same intermediate states $|\gamma\rangle$ occur in both expansions. But because of (31) we can conclude that $G_0(z)$ is single-valued at x_0 since x_0 is a pole of $D_0(z)$. This requires (v.H1) that none of the intermediate states $|\gamma\rangle$ be such that $D_\gamma(z)$ would have a cut extending through x_0. In other words $x_0 \leqslant B$. Hence the matrix element (90) and the operator $A_0(z)$ are single-valued at x_0. Hence

$$|\Psi_0\rangle^+ = |\Psi_0\rangle^- = |\Psi_0\rangle = O_0|\Phi_0\rangle \quad (4.91)$$

$$O_0 = -N_0^{-\frac{1}{2}}\mathscr{R}_{x_0}[A_0(z)]. \quad (4.92)$$

Applying considerations of previous sections to (89), we can write

$$A_0(E_0 + z) = \bar{A}_0(E_0 + z) * D_0(E_0 + z) \quad (4.93)$$

where

$$\bar{A}_0(E_0 + z) = -\frac{1}{z} + \int_\beta' \langle\beta|\bar{R}(E_0 + z)|\Phi_0\rangle \beta_{\text{op}}. \quad (4.94)$$

The matrix element $\langle\beta|\bar{R}(E_0 + z)|\Phi_0\rangle$ was defined earlier as the sum of all diagrams which have no ground state components. In terms of irreducible diagrams we have

$$\langle\beta|\bar{R}(E_0 + z)|\Phi_0\rangle = \langle\beta|[-\bar{D}(E_0 + z)V$$
$$+ \bar{D}(E_0 + z)V\bar{D}(E_0 + z)V - \ldots]_{\text{IL}}|\Phi_0\rangle(-z)^{-1}. \quad (4.95)$$

The important thing is the appearance of the factor $(-z)^{-1}$ in (95), which shows that $\bar{A}_0(E_0 + z)$ and $\langle\beta|\bar{R}(E_0 + z)|\Phi_0\rangle$ both have poles at the origin. Since $D_0(E_0 + z)$ has a pole at $z = x_0 - E_0$, and $A_0(E_0 + z)$

is given by (93), we have the following results: (a) $A_0(E_0 + z)$ has a pole at $x_0 - E_0$ and (b)

$$\mathscr{R}_{x_0}[A_0(z)] = N_0\mathscr{R}_0[\bar{A}(E_0 + z)] \equiv -N_0\bar{O}_0 \tag{4.96}$$

so that

$$O_0 = N_0^{\frac{1}{2}}\bar{O}_0. \tag{4.97}$$

Here the simplification arises from the fact that $\bar{O}_0$ involves only the diagrams without the ground state components. The equivalence of (97) to the Goldstone expression is easily demonstrated.

Further simplification in the operator O is possible when we realise that the diagrams for $\bar{A}(E_0 + z)$, *although linked are not in general connected.* They can have components of the type shown in Fig. 23. According to the Goldstone definition these are also linked in as much as they do not contain any 'unlinked' parts. This suggests the definition of a new operator,

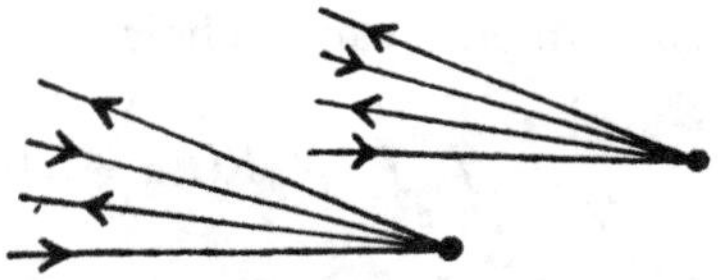

Fig. 23. An unconnected diagram which, according to Goldstone's definition, would be considered 'linked'.

$$\tilde{A}_0(E_0 + z) = \int_\beta' \langle\beta|\tilde{R}(E_0 + z)|\Phi_0\rangle\beta_{\text{op}} \tag{4.98}$$

where $\langle\beta|\tilde{R}(E_0 + z)|\Phi_0\rangle$ is the sum of all connected diagrams contributing to $\langle\beta|R(E_0 + z)|\Phi_0\rangle$ with $|\beta\rangle \neq |\Phi_0\rangle$. It follows that

$$\langle\beta|\tilde{R}(E_0 + z)|\Phi_0\rangle$$
$$= \bar{D}_\beta(E_0 + z)\langle\beta|[-V + V\bar{D}(E_0 + z)V - \ldots]_{\text{IC}}|\Phi_0\rangle(-z)^{-1}. \tag{4.99}$$

We now define

$$\tilde{O}_0 = -\mathscr{R}_0[\tilde{A}_0(E_0 + z)] \tag{4.100}$$

since for O_0 and $\bar{O}_0$ the residue is unique. More explicitly

$$\tilde{O}_0 = \int_\beta' \langle\beta|[-\bar{D}(E_0)V + \bar{D}(E_0)V\bar{D}(E_0)V - \ldots]_{\text{IC}}|\Phi_0\rangle\beta_{\text{op}}. \tag{4.101}$$

To see how to express $\bar{O}_0$ in terms of $\tilde{O}_0$ define

$$\bar{A}_0(E_0 + z) = \sum_{\nu=0}^{\infty} a_\nu \tag{4.102}$$

where a_ν is the sum of the contributions of all diagrams of $\bar{A}_0(E_0 + z)$ having ν components. It follows immediately from (98) and (94) that

$$a_0 = -z^{-1};\ a_1 = \tilde{A}_0(E_0 + z).$$

The second relation follows from the definition of $\tilde{A}_0(E_0 + z)$, since all diagrams contributing to the latter are connected they all have one component each. To find higher a_ν, consider all diagrams which are composed of a diagram of $\langle\beta'|\tilde{R}(E_0 + z)|\Phi_0\rangle$ and a diagram of $\langle\beta''|\tilde{R}(E_0 + z)|\Phi_0\rangle$. These diagrams are unconnected and have two components each. Their contribution to a_2 is given by

$$\mathop{\int'}_{\beta'} \mathop{\int'}_{\beta''} \langle\beta'|\tilde{R}(E_0 + z)|\Phi_0\rangle * \langle\beta''|\tilde{R}(E_0 + z)|\Phi_0\rangle \beta'_{\text{op}}\beta''_{\text{op}}.$$

Summing for all $\beta', \beta'' \neq \Phi_0$ we count each contribution to a_2 twice and hence

$$a_2 = \tfrac{1}{2}\tilde{A}_0(E_0 + z) * \tilde{A}_0(E_0 + z).$$

The diagrams of a_3 can be formed by combining a diagram of a_2 with a diagram of a_1 and so on

$$a_n = \frac{1}{n!}\underbrace{\tilde{A}_0(E_0 + z) * \tilde{A}_0(E_0 + z) * \ldots * \tilde{A}_0(E_0 + z)}_{n\ \text{factors}}.$$

Taking the residues of each term of (102) we have

$$\begin{aligned}\bar{O}_0 &= 1 + \tilde{O}_0 + \frac{1}{2!}\tilde{O}_0^2 + \frac{1}{3!}\tilde{O}_0^3 + \ldots\\ &= \exp(\tilde{O}_0).\end{aligned} \tag{4.103}$$

Finally

$$O_0 = N_0^{\frac{1}{2}} \exp(\tilde{O}_0) = \exp(-\tfrac{1}{2}\bar{G}_0'(E_0) + \tilde{O}_0). \tag{4.104}$$

These expressions are valid for a very large but finite volume Ω. The operator $\bar{O}_0$ involves diagrams without ground state components, whereas the operator $\tilde{O}_0$ involves only connected ground state diagrams and is therefore proportional to Ω as $\Omega \to \infty$. It follows that the norm of $\bar{O}_0|\Phi_0\rangle$ is exponentially large for $\Omega \to \infty$. However, in the same limit, in the norm of $O_0|\Phi_0\rangle$ compensation occurs since $\bar{G}_0'(E_0)$ is also proportional to Ω.

4f (ii). EXCITED STATE WAVE FUNCTIONS

We have now to discuss the operator $O^{\pm}_{\alpha}$. We start by observing that in general the matrix element $\langle\beta|R(E_0+z)|\alpha\rangle$ has diagrams having one or more components with external lines to the right. It can be written as a convolution of two matrix elements: $\langle\beta'|R(E_0+z)|\Phi_0\rangle$ which has no components with external lines to the right and the elements $\langle\beta''|\hat{R}(E_0+z)|\alpha\rangle$ which is the sum of the contributions of all those diagrams of $\langle\beta''|R(E_0+z)|\alpha\rangle$, all components of which have one or more external line to the right. We have

$$\langle\beta'\beta''|R(E_0+z)|\alpha\rangle = \langle\beta'|R(E_0+z)|\Phi_0\rangle * \langle\beta''|\hat{R}(E_0+z)|\alpha\rangle, \tag{4.105}$$

where

$$|\beta'\beta''\rangle = \beta'_{\text{op}}\beta''_{\text{op}}|\Phi_0\rangle,$$

and we have

$$\begin{aligned} A_\alpha(E_0+z) &= \int_{\beta'\beta''} \langle\beta'\beta''|R(E_0+z)|\alpha\rangle\beta'_{\text{op}}\beta''_{\text{op}} \\ &= \int_{\beta'\beta''} \langle\beta'|R(E_0+z)|\Phi_0\rangle * \langle\beta''|\hat{R}(E_0+z)|\alpha\rangle\beta'_{\text{op}}\beta''_{\text{op}}. \end{aligned} \tag{4.106}$$

On introducing the definition

$$\hat{A}_\alpha(E_0+z) = \int_\beta \langle\beta|\hat{R}(E_0+z)|\alpha\rangle\beta_{\text{op}} \tag{4.107}$$

one has

$$A_\alpha(E_0+z) = \hat{A}_\alpha(E_0+z) * A_0(E_0+z). \tag{4.108}$$

In the definition of $\langle\beta|\hat{R}(E_0+z)|\alpha\rangle$ the diagrams with ground state components are excluded [24]; application of earlier methods allows us to write

$$\begin{aligned} \langle\beta|\hat{R}(E_0+z)|\alpha\rangle = \langle\beta|[1 + \{-\bar{D}(E_0+z)V \\ + \bar{D}(E_0+z)V\bar{D}(E_0+z)V - \ldots\}_{\text{indR}}]|\alpha\rangle\bar{D}_\alpha(E_0+z) \end{aligned} \tag{4.110}$$

[24] The simplification has occurred in as much as $\hat{A}_\alpha$ is defined by means of diagrams having no ground state components and A_0 has been previously studied.

where **indR** implies that only the irreducible non-diagonal diagrams having one or more external lines to the right are to be considered.

The factor $\bar{D}_\alpha(E_0 + z)$ and, therefore, $\langle\beta|\hat{R}(E_0 + \bar{E}_\alpha)|\alpha\rangle$ has a pole[25] at $\bar{E}_\alpha$. The other factor on the right hand side of (110) has a cut along the *real* axis and is in most cases double valued at $\bar{E}_\alpha$. This situation was also encountered before and proceeding in an identical manner the two residues of (108) are

$$\mathscr{R}^{\pm}_{x_\alpha}[A_\alpha(z)] = \mathscr{R}^{\pm}_{x_\alpha}[\hat{A}_\alpha(E_0 + z)]\mathscr{R}_{x_0}[A_0(z)]. \tag{4.111}$$

Defining the operators

$$\hat{O}^{\pm}_\alpha = -\bar{N}_\alpha^{-\frac{1}{2}}\mathscr{R}^{\pm}_{x_\alpha}[\hat{A}_\alpha(E_0 + z)] \tag{4.112}$$

where $-\bar{N}_\alpha$ is the residue of $\bar{D}_\alpha(E_0 + z)$ in $\bar{E}_\alpha$, one gets

$$O^{\pm}_\alpha = \hat{O}^{\pm}_\alpha O_0 \tag{4.113}$$

and (113) applied to Φ_0 gives (85).

This completes the set of quantities needed to explicitly describe the perturbed system in its ground and excited states. It was assumed that $D_\alpha(z)$ has a pole. Actually, this has to be investigated for each particular problem. It turns out that for an interacting Fermi gas only for the lowest state $|\Phi_0\rangle$ has D_0 a pole; however, some of the low lying excited states also satisfy the criteria to a good approximation. On the other hand, in field theory all states $|\alpha\rangle$ have this property.

4f (iii). Some special states

One particle states: These are the states having one extra particle outside the Fermi sea. Let such a state be denoted by $|k\rangle$ where k is the momentum of the single particle. Then $\langle\beta|\hat{R}(E_0 + z)|k\rangle$ is formed from diagrams which are connected and have one external line at the right end. Now $\bar{D}_k(E_0 + z)$ are assumed to have a pole at $\bar{E}_k$ which requires that $\bar{G}_k(E_0 + z)$ be single valued at $\bar{E}_k$, i.e. the cut of $\bar{G}_k(E_0 + z)$ does not go through $\bar{E}_k$. This implies (v.H1) that the same property holds for all $\bar{D}_\gamma(E_0 + z)$ with $|\gamma\rangle$ occurring as an intermediate state in expansion (74) for $\bar{G}_\alpha(E_0 + z)$. Since the same inter-

[25] This is an assumption.

mediate states occur in the expansion for $\langle\beta|\hat{R}(E_0 + z)|k\rangle$ it follows that both this matrix element and the operator $\hat{A}_\alpha(E_0 + z)$ are single-valued at $\bar{E}_k$. Thus it is shown that for each single particle state $|k\rangle$, such that $D_k(E_0 + z)$ has a pole, the operators $\hat{O}_k^+$ and $\hat{O}_k^-$ are the same.

$$\hat{O}_k^+ = \hat{O}_k^- = \hat{O}_k \tag{4.114}$$

and we have only one stationary state

$$|\Psi_k\rangle^+ = |\Psi_k\rangle^- = |\Psi_k\rangle = \hat{O}_k|\Phi_0\rangle. \tag{4.115}$$

If there is more than one particle or hole present then $|\Psi\rangle^- \not\equiv |\Psi\rangle^+$. (See ref. v.H1 for a wave packet treatment of such one particle states.)

Asymptotically stationary states [26]:

In the field theory we can have states in which a number of "dressed" particles are present, each sufficiently far away from the other so that there is no interaction between them. Such states are stationary only in a limiting sense, i.e. they satisfy the Schrödinger equation only as $t \to \infty$. They are defined as

$$|\Psi_\alpha\rangle^{\mathrm{as}} = \hat{O}_{k_1}\hat{O}_{k_2} \ldots \hat{O}_{k_p}|\Psi_0\rangle \tag{4.116}$$

where

$$|\alpha\rangle = \xi_{k_1}^* \xi_{k_2}^* \ldots \xi_{k_p}^* |\Phi_0\rangle.$$

The wave function defined as a linear combination of (116)

$$|\Psi(t)\rangle^{\mathrm{as}} = \int_\alpha C_\alpha \, \mathrm{e}^{-\mathrm{i}x_\alpha t} \, |\Psi_\alpha\rangle^{\mathrm{as}} \tag{4.117}$$

does not satisfy the Schrödinger equation. However, the following wave function does so:

$$|\Psi(t)\rangle^{\pm} = \int_\alpha C_\alpha \, \mathrm{e}^{-\mathrm{i}x_\alpha t} \, |\Psi_\alpha\rangle^{\pm}. \tag{4.118}$$

The asymptotic nature of (116) is seen in the following limits

$$\lim_{t\to-\infty} \,||\Psi(t)\rangle^{\mathrm{as}} - |\Psi(t)\rangle^{+}| = 0$$

$$\lim_{t\to+\infty} \,||\Psi(t)\rangle^{\mathrm{as}} - |\Psi(t)\rangle^{-}| = 0.$$

In terms of diagrams

$$|\Psi_\alpha\rangle^{\mathrm{as}} = -N_\alpha^{-\frac{1}{2}} \mathscr{R}_{x_\alpha - E_0}[\oint_{\beta_0\beta_1\ldots\beta_p} |\beta_0\beta_1 \ldots \beta_p\rangle \langle\beta_0|R(E_0 + z)|\Phi_0\rangle$$

$$* \langle\beta_1|\hat{R}(E_0 + z)|k_1\rangle * \ldots * \langle\beta_p|\hat{R}(E_0 + z)|k_p\rangle]. \tag{4.119}$$

[26] See also ref. v.H1 where $|\Psi_\alpha\rangle^{\mathrm{as}} \equiv |\alpha\rangle_{\mathrm{as}}$.

The right hand side of (119) represents the sum over those diagrams of $\langle\beta|R(z)|\alpha\rangle$ each component of which has at most one external line at the ends. These are called *completely disconnected* diagrams. Taking the residues in x_α, we have

$$|\Psi_\alpha\rangle^{\mathrm{as}} = +N_\alpha^{\frac{1}{2}}[1 - D(x_\alpha)V + D(x_\alpha)VD(x_\alpha)V - \ldots]_{\mathbf{iD}}|\alpha\rangle \qquad (4.120)$$

iD $\equiv$ irreducible non-diagonal completely disconnected diagrams.

Metastable single particle states: In the foregoing it was found that if $D_\alpha(z)$ has a pole on the real axis then there exist stationary states $|\Psi_\alpha\rangle^\pm$ of the system defined by (47) and (83). On the other hand there are many systems which possess metastable states. As a matter of fact all states of a Fermi gas, with the exception of the ground state, should be metastable. Related problems occur in field theory of elementary unstable particles. The field theoretic problem has been discussed by Matthews and Salam (M6), by Schwinger (S7), by Van Hove (v.H1), and by Khalfin (K11).

The theory of metastable states of nuclei is not fully understood as yet. The first discussion attempted by Hugenholtz has been found to be unsatisfactory because of the difficulties with normalization (H4, N4). Recently Nosanow (N4) has improved the treatment in this respect and in connection with the isolation of a small parameter in the problem when the coupling constant itself is large.

However, one has a slightly better understanding of the most interesting special case of metastable states in nuclear physics. This is the case where a single particle, with momentum very close to the Fermi surface, moves through the nuclear medium. The particle retains its identity and the system may be pictured as an independent particle moving in an effective potential followed after some time by the absorption of the particle. The absorption is identified with the decay of the metastable state. It is assumed that the state decays into more complicated motions of the system. Sometimes one also says that the metastable state decays into collective motions. It should be emphasised that these statements do *not follow* from the phenomenological optical model or from the theory to be sketched below. They are rather intuitive extra statements which are needed to interpret the results of the model and the theory. A complete theory must, of course, describe the mechanism of interchange of energy between various states and it should also describe the state after the decay has taken place.

With these words of caution we go over to a description of metastable states in this special case.

Consider the addition of a particle of momentum k, $|k| > k_{\mathrm{F}}$, to the original system of N particles. The unperturbed state of this problem will be $|k\rangle = \xi_k^* |\Phi_0\rangle$. One can construct the operators $D_k(z)$ and $G_k(z)$. Let $G_k(z)$ be discontinuous along a cut on the real axis from E_{F} to $+\infty$ with a discontinuity $2\mathrm{i}J_k(x) \neq 0$ for z crossing the real axis at x. It follows that $D_k(z)$ has no pole from E_{F} to $+\infty$. But if $J_k(x)$ is sufficiently small near the root x_k of $E_k - x - K_k(x) = 0$ then one may write for x close to x_k

$$[D_k(x + \mathrm{i}0)]^{-1} = -\mathrm{i}J_k(x_k) + (x_k - x)[1 + K_k'(x_k) + \mathrm{i}J_k'(x_k)]. \quad (4.121)$$

It follows that

$$D_k(x + \mathrm{i}0) = \frac{N_k}{x_k - x - \mathrm{i}\Gamma_k}, \quad (4.122)$$

where

$$N_k = (1 + K_k'(x_k))^{-1}; \; \Gamma_k = N_k J_k(x_k). \quad (4.123)$$

For Γ_k very small, $D_k(z)$ behaves very nearly as if there were a pole at $z = x_k$ and may be defined by analytic continuation of (122) in z-plane near x_k. It is possible now to define a wave function $|\Psi_k\rangle^+$ which is approximately equal to the one given by (47). To display its metastable character one may follow the analysis indicated by (46) and (46′) to get

$${}^+\langle\Psi_{k'}|U(t)|\Psi_k\rangle^+ = \delta(k' - k) \exp(-\mathrm{i}x_k - \Gamma_k t) \quad (4.124)$$

which represents a state of approximate energy x_k which decays into other (unspecified) states after a time Γ_k^{-1}. The state $|\Psi_k\rangle^-$ may be defined corresponding to the other residue $D_k(x - \mathrm{i}0)$ it will have $\exp(+\Gamma_k t)$ time dependence and hence no physical significance may be attached to it.

Clearly it must be shown that $D_k(z)$ has the required property leading to (121). This depends principally on the momentum k and the nature of interactions in the system. A full discussion requires the solution of the set of coupled integral equations represented by [27]

$$\delta(k'-k)G_k(z) = \langle k'|[-V+VD(z)V-\dots]_{\text{idC}}|k\rangle$$
$$D(z) = [E-z-G(z)]^{-1}. \tag{4.125}$$

In practice this is very nearly an impossible task. By assuming reasonable properties for the spectrum x_k one may break the vicious circle of these equations and find answer to the question: what should be the nature of the spectrum x_k so that it may be identified with that of metastable states? Part of the answer is provided by Hugenholtz who observes that if $\bar{E}_F$ is the single particle energy corresponding to $k = k_F$ then the spectrum should be such that $\bar{E}_k > \bar{E}_F$ for $|k| > k_F$: $(\bar{E}_k = x_k - x_0)$. One may refer to the original paper (H3) for some approximate evaluations of the above expression. It must be remembered, however, that there is at present no way of knowing the extent of error involved in making these approximations. The question of metastable states or the connection between the optical model and the general many body problem as developed so far will not be discussed any further.

4g. *A theorem on single particle energies*

Although the situation with respect to metastable single particle states, as discussed in the previous section, is not very satisfactory, one can prove an important theorem about the single particle energies. This theorem says that for saturating systems the single particle energy at the top of the Fermi sea is equal to the average binding energy per particle. It finds practical use in providing a check on the validity of various approximate methods of calculating energies of the many body systems.

The energy of the state $|\Psi_k\rangle$ is obtained by looking at the singularities of the function $D_k(z)$. It is convenient to work instead with the function $\bar{D}_k(z)$ defined in Eq. (72),

$$D_k(z') = \bar{D}_k(z) * D_0(z'),\ z' = E_0 + z \tag{4.72}$$

where $\bar{D}_k(z')$ contains contributions only from the connected diagrams

[27] Γ_k is roughly equal to the imaginary part of the optical model potential; it is formally given by Eq. (123), (125). (See however (S8).)

having one particle line at both ends. As is clear from the nature of convolutions the poles of $\bar{D}_k(z')$ occur at the single particle energies $\bar{E}_k(= E_k - G_k(x_k) - E_0)$ (see the discussion of (75)).

To study this more carefully we introduce an operator $B(z)$ which is the sum of the contributions of connected diagrams to the matrix element of $R(z)$. That is,

$$B_\alpha(z) = \langle\alpha| \Big[- \frac{1}{H_0 - E_0 - z} V \frac{1}{H_0 - E_0 - z} + \frac{1}{H_0 - E_0 - z} V \frac{1}{H_0 - E_0 - z} V \frac{1}{H_0 - E_0 - z} \cdots \Big]_C |\alpha\rangle. \quad (4.126)$$

The contributions to an arbitrary matrix element $\langle\alpha|R(E_0 + z)|\alpha\rangle$ may be classified according to the number of their connected parts. Thus there are connected diagrams $(B(z))$, diagrams having two connected parts $\left[\frac{1}{2!} B(z) * B(z)\right]$, diagrams having three connected parts $\left[\frac{1}{3!} B(z) * B(z) * B(z)\right]$ and so on. Summing all these contributions, one has

$$\langle\alpha|R(E_\alpha + z)|\alpha\rangle = -z^{-1} + B(z) + \tfrac{1}{2}B(z) * B(z) + \frac{1}{3!} B(z) * B(z) * B(z) + \ldots \quad (4.127)$$

Recall that E_α is the energy of the unperturbed state. Hence $E_k = E_0 + k^2/2m$, so that

$$\langle k|R(E_k + z)|k\rangle = -z^{-1} + B_k(z) + \tfrac{1}{2}B_k(z) * B_k(z) + \ldots . \quad (4.128)$$

Define

$$\bar{B}_k(z) = B_k(z) - B_0(z). \quad (4.129)$$

The expansion of $D_0(E_0 + z)$ from (127) is

$$D_0(E_0 + z) = -z^{-1} + B_0(z) + \frac{1}{2!} B_0(z) * B_0(z) + \frac{1}{3!} B_0(z) * B_0(z) * B_0(z) + \ldots . \quad (4.130a)$$

Then using the equation (129) we have [28]

$$D_k(E_k + z)$$
$$= D_0(E_0 + z) * [-z^{-1} + \bar{B}_k(z) + \tfrac{1}{2}\bar{B}_k(z) * \bar{B}_k(z) \ldots]. \quad (4.130b)$$

Hence from (72)

$$\bar{D}_k(E_k + z) = -z^{-1} + \bar{B}_k(z) + \tfrac{1}{2}\bar{B}_k(z) * \bar{B}_k(z)$$
$$+ \frac{1}{3!}\bar{B}_k(z) * \bar{B}_k(z) * \bar{B}_k(z) \ldots . \quad (4.130c)$$

By definition $B_0(z)$ contains only connected ground state diagrams. In the limit $\Omega \to \infty$ $B_0(z)$ is, therefore, just proportional to Ω. The function $B_k(z)$ is defined to be the sum of contributions of all connected diagrams having only one particle line at both ends. As explained before these contributions differ from the corresponding ones in $B_0(z)$ only in having one factor $\delta(k_i - k)$ and one integration $\int_{k_i}$ less. In other words a diagram of $B_k(z)$ may be obtained from one of $B_0(z)$ by 'opening' one of the particle or hole lines (Fig. 24).

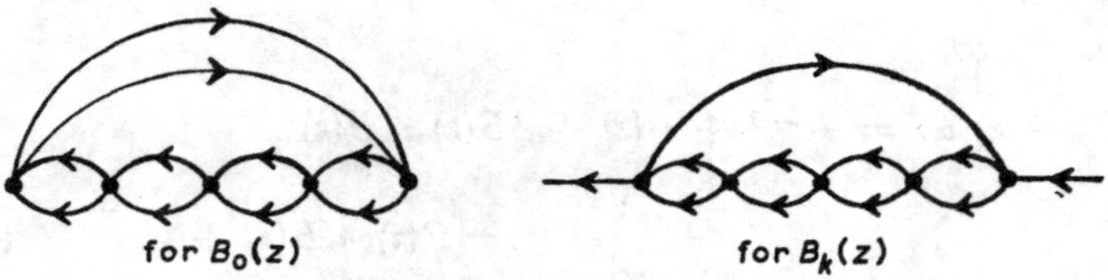

Fig. 24. Connection between diagrams of $B_0(z)$ and $B_k(z)$.

Mathematically the contribution to $B_k(z)$ from any diagram is obtained by replacing each $\int_{k_i}$ for particle lines in the corresponding contribution of $B_0(z)$ by $(\int_{k_i} - \Omega^{-1}(2\pi)^3 \times$ the same term with $k_i = k)$, and each $\int_{m_i}$ for hole lines by $(\int_{m_i} + \Omega^{-1}(2\pi)^3 \times$ the same term with $m_i = k)$. Thus it is seen that $B_k(z)$ has a part proportional to Ω which is exactly equal to B_0 and a part independent of Ω. The definition $\bar{B}_k = B_k - B_0$ thus makes $\bar{B}_k$ independent of Ω in the limit $\Omega \to \infty$. (This is the point at which the limit $\Omega \to \infty$ is taken.) In other words, to calculate $\bar{B}_k(z)$ first write $(2\pi)^3\Omega^{-1}B_0(z)$ as an integral over

[28] This is strictly true only for finite systems but transition to the infinite medium case ($\Omega \to \infty$) can be made afterwards without difficulty.

k_i's and m_i's then put each momentum in turn equal to k, with minus sign for $k_i = k$, integrate over all other momenta and sum such contributions.

Now notice that $B_0(z)/\Omega$ is a function of k_F which is the lower limit of integration of m_i and upper limit for k_i. Therefore, it follows that

$$\bar{B}_{k_F}(z) = 2\pi^2 k_F^{-2} \frac{d}{dk_F}\left(\frac{B_0}{\Omega}\right)$$

where the factor $(4\pi k_F^2)^{-1} 8\pi^3$ is required to complete the prescription for finding $\bar{B}(z)$. Since $\rho = k_F^3/6\pi^2$

$$\bar{B}_{k_F}(z) = \frac{d}{d\rho}(B_0(z)/\Omega). \tag{4.131}$$

From the definitions, Eqs. (126) and (74),

$$z^2 B_0(z) = \bar{G}_0(E_0 + z).$$

Now note that $\bar{D}_0(E_0 + z)$ has simple poles at $z = -\bar{G}_0(E_0)$ with residue given by $\exp(-\bar{G}_0'(E_0))$, equations (66) and (67). But $\bar{D}_k(E_k + z)$ is the same function of $z^2\bar{B}_k(z)$ that $\bar{D}_0(E_0 + z)$ is of $z^2 B_0(z)$ hence $\bar{D}_k(E_k + z)$ will have simple poles at

$$z = -\lim_{z_1 \to 0} [z_1^2 \bar{B}_F(z_1)] = -\frac{d}{d\rho}[\bar{G}_0(E_0)/\Omega]$$

with the residue

$$\exp\left[-\frac{d}{d\rho}(G_0'(E_0))/\Omega\right].$$

The pole of $\bar{D}_{k_F}(E_{k_F} + z)$ occurs at $\Delta E_F = E_F - k^2/2M$ where $E_F = \bar{E}_{k_F}$ is the Fermi energy. Hence

$$\Delta E_F = \frac{d}{d\rho}\left(\frac{\Delta E_0}{\Omega}\right).$$

This is a relation between the potential energy parts of E_F and the total energy $\mathscr{E}_0$. The same holds for the kinetic parts, and one has

$$E_F = \frac{d}{d\rho}\left(\frac{\mathscr{E}_0}{\Omega}\right) \tag{4.132a}$$

or

$$E_F = \left(\frac{\partial \mathscr{E}_0}{\partial N}\right)_\Omega \tag{4.132b}$$

or, since $N/\rho = \Omega$

$$E_F = \frac{\mathscr{E}_0}{N} + \rho \frac{\mathrm{d}}{\mathrm{d}\rho}\left(\frac{\mathscr{E}_0}{N}\right) \tag{4.132c}$$

or

$$E_F = \mathscr{E}_0/N + p/\rho \tag{4.132d}$$

where

$$p = -\left(\frac{\partial \mathscr{E}_0}{\partial \Omega}\right)_N = \rho^2 \frac{\mathrm{d}}{\mathrm{d}\rho}\left(\frac{\mathscr{E}_0}{N}\right)$$

is the pressure in the system. The above equations are the various equivalent statements of the important theorem of Hugenholtz and Van Hove (H4, K1, K16, V1). For *systems in equilibrium* the pressure vanishes and one has the result *that Fermi energy is equal to the average energy per particle.*

This theorem is rigorously true provided the perturbation expansions used here are valid and the D functions have the stipulated properties. The theorem is remarkable for not being dependent on the precise nature of the interactions.

It should be noted that the crux of the problem here is that E_F is defined by means of a certain type of diagrams while $\mathscr{E}_0$ is defined through other types of diagrams.

Although the beginning steps of the derivations, up to the definition of $\bar{B}_k$, have to be carried out for finite systems the theorem as finally stated in (132) is valid only for large systems. Presence of Ω^{-m} contributions may be of significance for finite nuclei, in which case the theorem will not be exactly valid.

As a last step in this discussion we show that the present formalism contemplates systems which have a continuous single particle energy spectrum for particles having momentum l in a small neighbourhood for the Fermi momentum k_F. Consider the function $\bar{D}_m(z)$ for holes ($|m| < k_F$) which is given by an analysis similar to that for $\dot{D}_k$.

$$\bar{D}_m\left(E_0 - \frac{m^2}{2M} + z\right) = -z^{-1} + \bar{B}_m(z) + \tfrac{1}{2}\bar{B}_m(z) * \bar{B}_m(z) + \frac{1}{3!}\bar{B}_m(z) * \bar{B}_m(z) * \bar{B}_m(z) + \dots .$$

In defining $\bar{B}_m(z)$, naturally, the role of particles and holes is interchanged. It is seen that for $|m| \to k_F$, $-\bar{B}_m(z) \to \bar{B}_{k_F}(z)$. On shifting the poles appropriately

$$\bar{D}_k(E_0 + z) * \bar{D}_m(E_0 + z) = -z^{-1} \text{ for } |m| = |k| = k_F.$$

The sum of the poles is zero, and product of the residues is one. It follows that the energy of a *hole* at the Fermi surface is equal to $-E_F$. Hence the energy E_l of a *particle* with momentum $|l|$ close to k_F is continuous at $|l| = k_F$. This is not necessarily the case, for instance, in the theory of superconductivity the single particle spectrum is not continuous but exhibits a so-called energy gap near the Fermi surface. The existence of this energy gap being crucial for the appearance of the phenomenon of superconductivity (B1, B14). Klein (K1, K16) has given a different proof of the theorem of Hugenholtz and Van Hove. In particular he has shown that the theorem is true only for the 'normal fluids' case (K16). We shall not be into a discussion of these theories, but a few brief remarks will go made at the end (Chapters IV and V).

LIST OF IMPORTANT SYMBOLS IN CHAPTER II

(*op. def.* means operator defined as the)

$\frac{1}{a}$	the propagator or the operator generating the energy denominators
a_k	arbitrary real number
a_ν	*op. def.* sum of the contributions of all diagrams of $\bar{A}_0$ having ν components, (4.102)
A	number of particles in system – same as N
$A(z)$	*op. def.* the operator whose residues determine the operator $O^{\pm}_{\alpha}$, and therefore the perturbed wave functions, (4.86)
$\bar{A}(z)$	*op. def.* sum of all diagrams of $A(z)$ having no ground state components, (4.93)
$\tilde{A}(z)$	*op. def.* sum of all connected diagrams of $A(z)$, (4.98)
$\hat{A}(z)$	*op. def.* sum of all diagrams of $A(z)$ which have one or more ex-

ternal lines to the right, (4.107, 4.108)

b_k arbitrary real numbers

$B(z)$ *op. def.* sum of all connected diagrams of $R(z)$, (4.126)

$\bar{B}(z)$ *op. def.* difference of $B_0(z)$ and $B(z)$. Independent of Ω in the limit $\Omega \to \infty$, (4.129)

$D(z)$ diagonal part of the resolvent operator or Green's function, (4.23), (4.29), (4.33)

$\bar{D}_\alpha(z)$ *op. def.* sum of diagrams of $D_\alpha(z)$ ($\alpha \neq 0$) which do not contain ground state components, hence independent of Ω in the limit, (4.73)

E_α energy of the unperturbed system corresponding to the state $|\alpha\rangle$ (denoted in Ch. I by K_i)

E'_α excitation energy of the state $|\alpha\rangle$ in the unperturbed system ($= E_\alpha - E_0$)

$\bar{E}_\alpha$ true excitation energy in the perturbed system ($= \mathscr{E}_\alpha - \mathscr{E}_0$)

E_{F} the Fermi-energy, the excitation energy of a single particle state with momentum k_{F}

$\mathscr{E}_\alpha$ energy of the perturbed system corresponding to the state $|\alpha\rangle$. Also denoted by x_α

$f(z)$ functions of z having the same analytic properties as the resolvent operator

$F(a_n \ldots a_0, b_m \ldots b_0)$ function of complex variable z having the same structure as that of contributions from diagrams, (4.50)

$F_{\alpha\beta}$ non diagonal part of the resolvent operator, (4.22)

$g(z)$ function of z having same analytic properties as $R(z)$

$G(z)$ operator that determines the level shift, (4.30, 4.32)

$\bar{G}(z)$ *op. def.* sum of all connected diagrams of $G(z)$. $\bar{G}_0$ is proportional to Ω, (4.55), while $\bar{G}_\alpha$ having no ground state components is independent of Ω, (4.74), in the limit

$G'_\alpha(x_\alpha)$ derivative of $G_\alpha(x_\alpha)$ at x_α

$h(z)$ *op. def.* by means of $\bar{G}_0$ connected to G'_0, (4.61)

H total or perturbed Hamiltonian

H_0 unperturbed Hamiltonian

$J_\alpha(x_\alpha)$ imaginary part of $G_\alpha(x_\alpha)$

$\bar{J}_\alpha(x_\alpha)$ imaginary part of $\bar{G}_\alpha(x_\alpha)$
$K_\alpha(x_\alpha)$ real part of $G_\alpha(x_\alpha)$
$\bar{K}_\alpha(x_\alpha)$ real part of $\bar{G}_\alpha(x_\alpha)$
N number of particles in the system (also denoted by A)
N_α negative of the residue of D_α in x_α, inverse square of the norm of the wave function, (4.44, 4.47)
$O^\pm_\alpha$ operator which generates the perturbed ingoing (—) and outgoing (+) stationary states $\Psi^\pm_\alpha$ from the unperturbed vacuum, (4.83)
$\bar{O}$ *op. def.* sum of diagrams of O having no ground state components i.e., defined through linked diagrams, (4.96)
$\tilde{O}$ *op. def.* sum of only connected diagrams of O hence proportional to Ω in the limit, (4.100)
$\hat{O}^\pm$ *op. def.* sum of the diagrams of O having one or more external lines to the right. Generates the perturbed in and outgoing stationary states from the *perturbed* vacuum, (4.85, 4.112)
$|0_A\rangle$ absolute vacuum having no particles in it
P time ordering operator, (3.33)
$\mathscr{R}^\pm_{x_\alpha}(\ldots)$ the two possible residues at x_α of the operator in the parentheses
$R(z)$ the resolvent operator. It is negative of the Green's function $W(E)$ of Ch. I when E is replaced by z
$\bar{R}(z)$ *op. def.* sum of all diagrams of $R(z)$ which have no ground state components, (4.53, 4.95)
$\tilde{R}(z)$ *op. def.* sum of all connected diagrams of $R(z)$
$\hat{R}(z)$ *op. def.* sum of all diagrams of R which have one or more external line to the right
T total kinetic energy
T_i single particle kinetic energy
$U(t)$ unitary transformation operator from Heisenberg to Schrödinger representation, (4.15), (4.17)
$U(t, t')$ unitary transformation operator which in the interaction representation takes the system from time t' to t
$U_\alpha(t, t')$ same as $U(t, t')$ but with provision for adiabatic switching off of the interaction in infinite past
U_i single particle potential energy
V the perturbation potential or interaction

v_{ij} the actual two-body potential, (2.4)

$\bar{v}_{ij}$ the 'effective' two-body potential which results on subtracting the effect of the single particle potential from v_{ij}, (2.5′)

$V(t)$ interaction representation of the perturbation, (3.30)

$v(klmn)$ the antisymmetrised form of the Fourier transform of the two-body interaction, (4.8, 4.9)

x_α energy of the perturbed system corresponding to state Ψ_α. The symbol is used to emphasise that the energy is only approximate (4.41), otherwise same as $\mathscr{E}_\alpha$

z an arbitrary complex number being a generalisation of the energy variable ($= x + iy$)

Γ_k inverse of the life time of a single particle state with momentum k

$\delta(\beta - \alpha)$ product of delta functions involving momenta of *all* the particles and holes in the states β and α. Occurs only if the two states have the same number of particles and holes

ϵ_i single particle energies

$\epsilon_{i,r}, \epsilon_{ij,rs}$ energy denominators of singly and doubly excited intermediate states, (2.13)

ζ complex variable in energy plane, alternative for z

η imaginary part of complex number

$\eta_i(\eta_i^*)$, $[\equiv \eta_{k_i}(\eta_{k_i}^*)]$ annihilation (creation) operator for particle in state k_i

ξ real part of the complex number ζ

$\xi_i(\xi_i^*)$, $[\equiv \xi_{k_i}(\xi_{k_i}^*)]$ annihilation (creation) operator for particle in state k_i with normalisation appropriate to a large system

ρ particle density in the system ($= N/\Omega$)

$\varphi(\xi)$ function of the variable ξ, not a wave function, (4.68)

Φ generic symbol for unperturbed wave functions

ψ_i, $(\equiv \psi_{k_i})$ single particle wave functions

$\psi(z)$ function of a complex variable, not to be confused with the wave function ψ_i above, (4.66)

$\psi(\boldsymbol{x}_i)$ field operator at the point $\boldsymbol{x}_i$

Ψ generic symbol for perturbed wave functions

$|\Psi_\alpha\rangle^\pm$ in ($-$) and out ($+$) going

	stationary states of the perturbed system corresponding to the unperturbed state $\vert\alpha\rangle$, (4.47)
$\vert\Psi_\alpha\rangle^{\mathrm{as}}$	asymptotically stationary state of perturbed system
$\Psi_{\mathrm{S}}(t)$	perturbed wave function at time t in the Schrödinger representation
$\Psi(t)$	perturbed wave function at time t in the interaction representation
Ω	volume of the system

Note on labeling of states:

Eigenstates of H_0 are labeled by Latin or Greek indices, thus: $|\Phi_n\rangle$, $|\Phi_\alpha\rangle$ or simply by $|\alpha\rangle$, *but not* $|n\rangle$. Single particle states or eigenstates of $(T_i + U_i)$ are represented by ψ_i or ψ_{k_i}. Common Latin indices are used to denote these states. Two different conventions are used for distinguishing the states in and outside the chosen configuration:

(a) With only one subscript or label we put a bar on top of the label if the state is inside the chosen configuration. (Used in section 2.)

(b) With two symbol indication all states k_i are outside the chosen configuration (particles) and all states m_i are inside (holes) an arbitrary state either in or outside the state is denoted by symbols n_i, l_i, etc. (Used in section 4.)

CHAPTER III

REARRANGEMENT METHODS: REACTION MATRIX

1. INTRODUCTION

For the previous discussion the validity of the perturbation expansion seems to be required. However, it is known that for the three most important interactions encountered in nature the usual perturbation expansions are as such not valid. In the case of nuclear matter the nucleon-nucleon potential has a hard core and each individual matrix element of V with respect to any single particle wave function is infinite. For an electron gas (Coulomb potential) the infinity arises from the low momentum transfer (long range) part of the interaction and the terms beyond first order are infinite (G10, U1). A still different type of singularity (B1) occurs in certain terms of the electron-phonon interaction which is responsible for superconductivity. In all these cases it has been shown that these infinities disappear if one sums up the contribution of certain appropriate classes of diagrams before performing the integrations over momentum variables. Thus it seems that the existence of the basic perturbation expansion is perhaps too narrow a criterion.

For the nuclear problem the so-called reaction matrix methods have been used to overcome these difficulties. These methods have their origin in the multiple scattering theories developed in the last decade by Watson and Brueckner. The application to the problem of calculating nuclear binding energies was made by Brueckner (B11, B12) and collaborators. In the first chapter we have given a concise derivation of the Brueckner method. The problem of anomalous A dependence of terms mentioned there is most satisfactorily solved by combining the t-matrix methods with the linked cluster theorem proved in the second chapter. The t-matrix scheme is fairly flexible because of the various possibilities of defining the propagator. A practical way of defining the propagator is through the introduction of single particle energies, thus making contact with the phenomenological, independent particle nuclear models.

The discussion has been arranged in order of increasing complexity and the important alternative points of view have been given.

2. VERTEX MODIFICATION: THE t-MATRIX

To improve upon the ordinary perturbation theory of the nuclear problem one has to sum the terms of the series in such a way that it is effectively equivalent to replacing each individual two particle matrix element of v, which is infinite or very large, by a finite and manageable quantity. The possibility of such a program is suggested by the fact that in spite of the hard core the nucleon-nucleon scattering amplitudes are finite quantities. If we look upon the nuclear binding as resulting from multiple scattering inside nuclear matter, we may hope to replace the infinite matrix elements of v_α everywhere by finite scattering amplitudes. This is accomplished by introducing a t-matrix as in Ch. I, sec. 6b through an integral equation involving v_α. Recall the definition

$$t_\alpha = v_\alpha \left[1 + \frac{1}{e} P t_\alpha \right]$$

$$e = E - K - O$$

where the operators P and O are arbitrary except that they must satisfy the equation

$$(1 - P) VM |p_0\rangle = OM |p_0\rangle$$

of Riesenfeld and Watson described earlier (Ch. I, sec. 6).

In view of this arbitrariness in the defining equation there occur a number of operators in the literature which differ slightly from each other but go under the name of reaction (or t-) matrix. (Bethe (B15) denotes this matrix by G, while in their later papers Brueckner *et al.* (B19) use the symbol K.) The important common feature is that the definition is made through an integral equation. The over-riding consideration is that the new operator t_α should have finite matrix elements between the basic free particle wave functions.

However, it is not necessary to appeal to multiple scattering. One can look upon the t-matrix equation as an auxiliary to the perturbation theoretic expansion. This is the point of view advocated by Tobocman (T5). He has a slightly different definition of the t-matrix

$$t_\alpha = v_\alpha + v_\alpha g_\alpha t_\alpha - \delta_\alpha v_\alpha g_\alpha v_\alpha^{-1} t_\alpha \tag{2.1}$$

where the arbitrariness or rather the flexibility of the definition resides in the so far unspecified propagator g_α and the constant number δ_α. The equation (1) can be inverted

$$\begin{aligned} v_\alpha &= t_\alpha(1 + g_\alpha t_\alpha)^{-1}(1 + g_\alpha \delta_\alpha) \\ &= (t_\alpha - t_\alpha g_\alpha t_\alpha + t_\alpha g_\alpha t_\alpha g_\alpha t_\alpha \ldots)(1 + g_\alpha \delta_\alpha). \end{aligned} \tag{2.2}$$

Recall now the R–S expansion for the level shift discussed previously

$$\begin{aligned} \Delta E = \mathscr{E} - E_0 = \langle V \rangle + \langle V \frac{1}{a} V \rangle + \langle V \frac{1}{a} V \frac{1}{a} V \rangle + \ldots \\ - \langle V \rangle \langle V \frac{1}{a} \frac{1}{a} V \rangle \ldots \end{aligned} \tag{2.3}$$

where

$$\begin{aligned} a^{-1} &= (E_0 - H_0)^{-1}(1 - \Lambda_0) \\ V &= \sum_\alpha v_\alpha - U = \sum_\alpha v_\alpha - \sum_i U_i \\ &\equiv \sum_\alpha (v_\alpha - u_\alpha) \\ u_\alpha &= u_{ij} = \tfrac{1}{2}(U_i + U_j)\,; \; \alpha = (i, j). \end{aligned}$$

Or, reducing in terms of the interaction

$$\begin{aligned} \Delta E &= \left\{ \langle (\Sigma\, v_\alpha - U) \rangle + \langle (\Sigma\, v_\alpha - U) \frac{1}{a} (\Sigma\, v_\alpha - U) \rangle + \ldots \right\} \\ &\quad - \left\{ \langle (\Sigma\, v_\alpha - U) \rangle \langle (\Sigma v_\alpha - U) \frac{1}{a} \frac{1}{a} (\Sigma\, v_\alpha - U) \rangle + \ldots \right\} \\ &\equiv A_{\mathrm{RS}} - B_{\mathrm{RS}} \end{aligned} \tag{2.4}$$

where A_{RS} and B_{RS} refer to the terms grouped in the first and second lines respectively.

The single particle potential is written explicitly here to emphasise that in the nuclear problem we do not *a priori* know the single particle potential U. From the success of shell model we are led to assume that it must be possible to make a choice of U such that the level shift ΔE is very small or vanishes. At the moment, however, it is assumed to have a certain given form and the interest is to see how the convergence of the series (4) can be improved. One can use (2) to eliminate

v_α in favour of t_α from (4). This gives the series

$$\Delta E = \left\{\langle \sum t_\alpha - U\rangle + \langle(\sum t_\alpha - U)\frac{1}{a}(\sum t_\alpha - U)\rangle + \ldots\right\}$$

$$- \left\{\langle \sum t_\alpha - U\rangle \langle(\sum t_\alpha - U)\frac{1}{a}\frac{1}{a}(\sum t_\alpha - U)\rangle + \ldots\right\}$$

$$- \left\{\langle \sum t_\alpha g_\alpha t_\alpha\rangle + \ldots\right\} \equiv A_t - B_t - C_t, \tag{2.5}$$

which has a new class of terms, C_t, grouped in the third line. Alternatively one might define another operator

$$\tau_\alpha = (v_\alpha - u_\alpha)[1 + g_\alpha(\tau_\alpha - \delta_\alpha[v_\alpha - u_\alpha]^{-1}\tau_\alpha)] \tag{2.6}$$

and eliminate both v_α and u_α from (4) to obtain an expansion in terms of a single quantity τ_α with obvious advantages for certain purposes.

$$\Delta E = \left\{\langle \sum \tau_\alpha\rangle + \langle(\sum \tau_\alpha)\frac{1}{a}(\sum \tau_\alpha)\rangle + \ldots\right\}$$

$$- \left\{\langle \sum \tau_\alpha\rangle \langle(\sum \tau_\alpha)\frac{1}{a}\frac{1}{a}(\sum \tau_\alpha)\rangle + \ldots\right\} - \left\{\langle \sum \tau_\alpha g_\alpha \tau_\alpha\rangle + \ldots\right\}$$

$$\equiv A_\tau - B_\tau - C_\tau. \tag{2.7}$$

The structure of the terms A and B is the same in all three equations (4), (5) and (7). As has been shown earlier A and B together represent the contributions from linked diagrams only. The difference comes in the operators to be used at each vertex (or interaction line in Goldstone graphs) viz., $(v_\alpha - u_\alpha)$, $(t_\alpha - u_\alpha)$, (τ_α). This changing of the series based on one interaction operator into a series based on a different interaction operator is often called vertex modification. The modification will be an improvement over (4) provided that the new terms brought in by C cause cancellations in A and B in an appropriate way. These terms depend on g_α and δ_α and the criterion for their choice is clearly that they induce maximum balanced cancellations. In view of the complicated terms in higher orders one has to be careful to ensure that the cancellations by C in A do not undo those by C in B. One wants cancellations in order that the number of terms that actually contribute is smaller than in the original expansion (4).

Two choices will be discussed. The more usual one is

$$g_\alpha = a^{-1}, \ \delta_\alpha = 0. \tag{2.8}$$

But Tobocman (T5) prefers the choice

$$g = a^{-1}, \ \delta_\alpha = \langle t_\alpha \rangle. \tag{2.9}$$

Considering the first choice (8) for g_α and δ_α. One has

$$C_\tau = - \langle \Big\{ \sum_\alpha \Big(\tau_\alpha \frac{1}{a} \tau_\alpha \Big) + \sum_\alpha \Big(\tau_\alpha \frac{1}{a} \tau_\alpha \frac{1}{a} \tau_\alpha \Big) + \dots \Big\} \rangle.$$

These terms occur in A_τ with opposite signs but as parts of the terms like

$$\langle (\Sigma\, \tau_\alpha) \frac{1}{a} (\Sigma\, \tau_\beta) \rangle, \quad \langle (\Sigma\, \tau_\alpha) \frac{1}{a} (\Sigma\, \tau_\beta) \frac{1}{a} (\Sigma\, \tau_\gamma) \rangle.$$

Thus the cancelled terms are those in which the indices of the *same pair* of particles occur consecutively. This cancellation proceeds to all orders. Comparing now the expansion in terms of v_α and τ_α (or t_α) one sees that the latter has lesser number of terms.

One also sees that

$$\langle t_\alpha \rangle = \langle v_\alpha \rangle + \langle v_\alpha \frac{1}{a} v_\alpha \rangle + \langle v_\alpha \frac{1}{a} v_\alpha \frac{1}{a} v_\alpha \rangle + \dots$$

so that, the cancellation refered to above, or the introduction of the t-matrix, is equivalent to a partial summation of the terms on the right hand side. This is usually expressed by saying that the replacement of v by t-matrix in perturbation theory takes into account the successive interactions between any pair of particles to all orders, or that the t-matrix exactly takes into account the pair interactions.

To consider the effect of the choice (9) for g_α and δ_α, take the third order contributions

$$A_\tau^{(3)} = \sum_{\alpha\beta\gamma} \langle \tau_\alpha \frac{1}{a} \tau_\beta \frac{1}{a} \tau_\gamma \rangle \tag{2.10a}$$

$$-B_\tau^{(3)} = \sum_{\alpha\beta\gamma} \langle \tau_\alpha \rangle \langle \tau_\beta \frac{1}{a} \frac{1}{a} \tau_\gamma \rangle \tag{2.10b}$$

$$-C_\tau^{(3)} = \sum_{\alpha\beta}\Bigg[\langle\left\{\left(\tau_\alpha\frac{1}{a}\tau_\alpha\frac{1}{a}\tau_\beta\right)+\left(\tau_\alpha\frac{1}{a}\tau_\beta\frac{1}{a}\tau_\beta\right)\right\}\rangle$$

$$+\langle\tau_\alpha\frac{1}{a}\tau_\alpha\frac{1}{a}\delta_\alpha\rangle \tag{2.10c}$$

$$-\left\{\langle\tau_\alpha\rangle\langle\tau_\alpha\frac{1}{a}\frac{1}{a}\tau_\beta\rangle+\langle\tau_\alpha\rangle\langle\tau_\beta\frac{1}{a}\frac{1}{a}\tau_\alpha\rangle\right\}\Bigg].$$

where the term in the second line vanishes.

The effect of δ_α ($=\langle\tau_\alpha\rangle$) terms in $C_\tau^{(3)}$ is to cancel terms in $B_\tau^{(3)}$ to make the restriction on the sum there $\alpha \neq \beta \neq \gamma$, just as the terms in the first line of $C_\tau^{(3)}$ cancel the terms in $A_\tau^{(3)}$ to impose the same restriction on the sum there. This is better than the choice $\delta_\alpha = 0$ of (8) because in that case only the terms in $A_\tau^{(3)}$ are cancelled and the balance between A and B terms that existed before the introduction of t-matrix is partly disturbed. The discussion for higher orders is similar.

In connection with Hugenholtz's method the partial summation by propagator modification was discussed. This means of additional partial summations is still open to us. The previous discussion goes through because no explicit definition of the propagator, a, was used there.

The present way of looking at the rearrangements is similar to the one followed by Thouless (T6) where it makes for simplification in the discussion of some points.

2a. *Graphical representation of the t-matrix equation*

In terms of Hugenholtz graphs the t-matrix equation can be represented as in Fig. 25a if we make the convention that a heavy dot at the vertex represents a t-interaction. This makes explicit a fact not clearly mentioned so far, viz., that the equation defining t takes into account only the particle-particle scattering terms Fig. 25a, b. It sums up the particle-ladders of the type shown in Fig. 25c.

One can also consider other types of ladders which include hole-hole, hole-particle and pair creation diagrams. But this will greatly complicate the integral equations which have to be solved. Inclusion of the hole-hole interactions has been studied by Mehta and by Chisholm and Squires (C2, E5, M3) and is shown to modify some details of the solution. More complex ladders have been studied by Mehta (M9). The possibility of defining other types of ladders arises from the possibility of making different partial summations in the

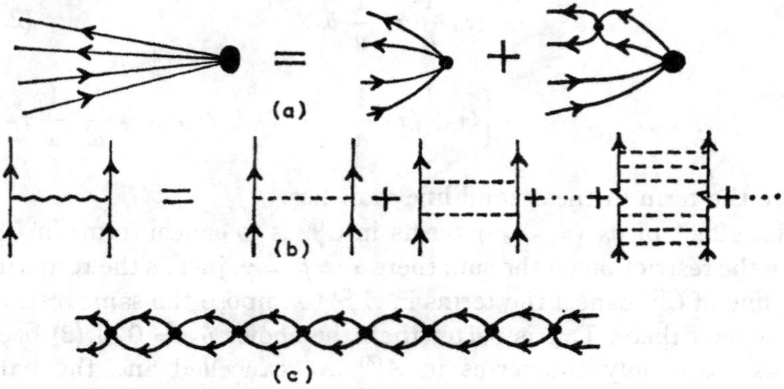

Fig. 25. Diagrammatic representation of t-matrix equation (a and b). Heavy dot in (a) and the wavy line in (b) indicate a t-vertex. The r.h.s. of (b) shows in Goldstone's representation the ladder diagrams obtained by iteration. (c) shows a typical particle ladder in Hugenholtz representation.

series. The formal development described by equation (1) to (10) is valid for any consistent definition of the t-matrix. The diagramatic way of writing the equations makes explicit the intermediate states which are to be included in solving for the t-matrix. Brueckner believes that in the neighbourhood of the usual nuclear density only the particle-particle interactions will predominate and in his work only the above restricted definition of the t-matrix has been used.

It should always be kept in mind that whatever the definition of t-matrix we employ, there will be in any practical calculation terms which will have to be left out in energy and t-matrix expressions: The problem of proving the method consists in demonstrating that the omitted terms are small. The same thing is often expressed by saying that one has to prove that the rearrangement of the series is justified.

3. PROPAGATOR MODIFICATION

The quantity $1/a$ occurring in previous expressions has been refered to as the propagator or the Green's function. It is the quantity arising from the energy denominators of the intermediate states. In the R–S perturbation theory, as we know, it is $1/a = (1 - \Lambda_{p_0})/(E_0 - H_0)$. This might be interpreted as saying that between any two interactions the particles 'move' [29] according to the zero-order Hamiltonian; hence the name propagator. That is, they are 'free' particles between any two interactions. A 'physical' argument is now given: As the interacting particles are inside the medium they are no longer governed by the zero-order Hamiltonian. One should therefore modify the propagator into such a way that the effect of the medium is more correctly taken in account between any two successive collisions. This type of argument is quite unnecessary and it may mislead one to think that something was wrong with R–S perturbation theory. This is not so. What one wants to do is the same as was done in (II. 4.27), (II. 4.28). There we found that if we replaced $(E_\alpha - z)^{-1} = \langle\alpha|(E_0 - H_0)^{-1}|\alpha\rangle$ by D_α everywhere in the expansion of $\langle\alpha|R(z)|\beta\rangle$, then we could reduce the number of diagrams (or terms) that need be considered. This reduction in the number of terms must mean an improvement of the convergence of the series. As in connection with the t-matrix, here also one uses the slang: 'such and such diagrams have been eliminated or removed'. What is meant is again as before: 'such and such diagrams have been summed up by the introduction of the t-matrix or by propagator modification, and their contributions are now included in other diagrams in virtue of the modification'. The replacement $1/a = (E_0 - H_0)^{-1} \to$ diagonal part of $(E_0 - H)^{-1}$ is the maximum modification that one can make. It leads to the maximum reduction in the number of terms. It is not always a practical thing to do, however.

To understand more fully the nature of this modification, consider the replacement

$$\frac{1}{a} = \frac{1 - \Lambda_0}{E_0 - H_0} \to \frac{1 - \Lambda_0}{E_0 + V_{ii}^0 + V_{jj}^0 - H_0 - V_{ii} - V_{jj}} = \frac{1}{b}$$

[29] Such descriptions are highly 'symbolic'. Unfortunately they sometimes can also be quite misleading. It is felt nonetheless that one must introduce the current terms and indicate their meaning, although it is entirely possible and perhaps also desirable to have a more precise terminology.

where

$$V_{ii} = \sum_{k \neq i} v_{ik}, \quad V_{ii}^0 = \langle 0 | V_{ii} | 0 \rangle.$$

On expanding the new propagator in the powers of

$$\Delta_v = (V_{ii}^0 + V_{jj}^0 - V_{ii} - V_{jj})$$

one gets

$$\frac{1}{b} = \frac{1}{a}\left(1 - \frac{\Delta_v}{a} + \frac{\Delta_v^2}{a^2} - \ldots\right).$$

Now in second order if we have

$$\langle V \frac{1}{b} V \rangle \text{ instead of } \langle V \frac{1}{a} V \rangle$$

we have not only the diagram of Fig. 26a, but also all the higher diagrams (up to infinite order) of the type shown Figs. 26b–d. . .

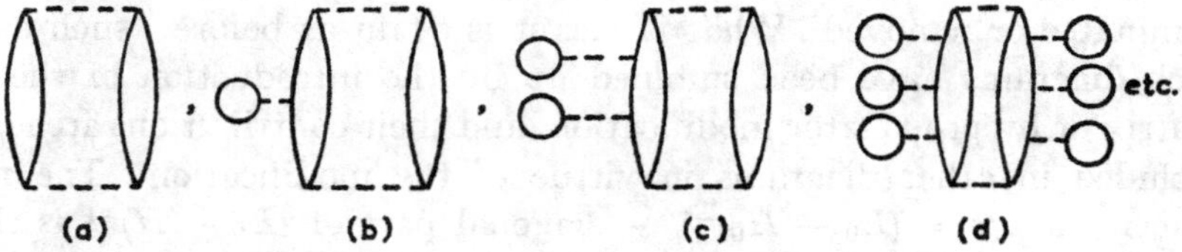

Fig. 26. Diagrams summed by a propagator modification.

These should therefore be omitted when we come to calculate higher orders with the propagator $1/b$ instead of $1/a$. The propagator used in practical applications of the Brueckner method is the following

$$\frac{1}{e} \equiv \frac{1 - \Lambda_0}{(E_0 + t_{ii}^{(0)} + t_{jj}^{(0)} - H_0 - t_{ii} - t_{jj})}$$

where

$$t_{ii} = \sum_{j \neq i} t_{ij}, \; t_{ii}^{(0)} = \langle 0 | t_{ii} | 0 \rangle.$$

The difference now is that at each vertex we have a t-interaction rather than a v-interaction but with the restriction that the state of the particle in the Fermi sea is not changed.

In considering examples of diagrams summed by this modification, recall that the second order v-term is included in the first order t-term together with other, higher order v-terms. Consider the diagrams starting from the third order shown in Fig. 27a. The present propagator $1/e$ then sums up the diagrams (which are themselves sums of many diagrams) of the type shown in Fig. 27b. The propagator modification is sometimes represented by changing the thickness or doubling the particle-hole lines.

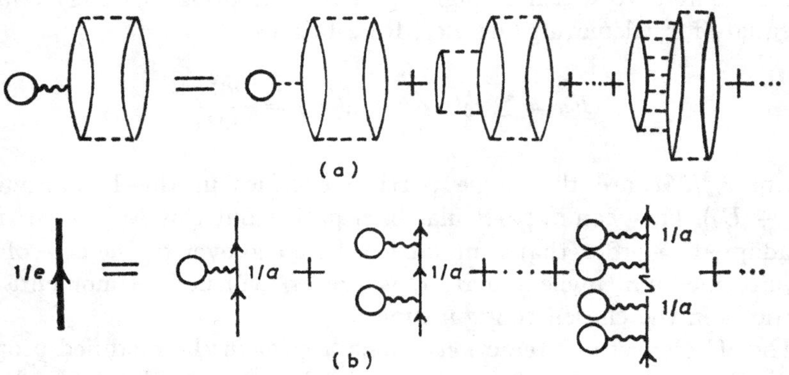

Fig. 27. Diagrams showing steps in a more complex vertex modification involving the propagator $1/e$.

The interaction with an unexcited particle of the chosen configuration which leaves its state unchanged is called 'forward scattering'. Hence the propagator modification is said to take into account the 'forward scattering' in the medium – by v-interaction for $1/b$ and by t-interaction for $1/e$. As mentioned before, maximum advantage is obtained by the replacement $1/a \to D$ which leaves only the irreducible diagrams to be calculated.

4. MODIFIED PROPAGATION IN THE t-MATRIX

The equation which we now write for the t-matrix is a non-linear integral equation

$$t_{ij} = v_{ij} + v_{ij} \frac{1 - \Lambda_0}{E_0 + t_{ii}^{(0)} + t_{jj}^{(0)} - H_0 - t_{ii} - t_{jj}} t_{ij}. \tag{4.1}$$

As has been hinted before we can expand the propagator and find out what terms are summed up in this case.

It is more useful to look at this problem from the point of view of propagation of single particles. The purpose of the theory is to see the extent to which a single particle description of the system is adequate for calculating energies. Recall that

$$E_0 = \sum_i \epsilon_0(m_i); \quad \epsilon_0(m_i) = \frac{m_i^2}{2M}$$

where $m_i^2/2M$ are the single particle energies in the Hamiltonian $(T_i + U_i)$. They can in particular be negative but this way of writing is adopted in order that it may be easier to go over to the case of an infinite medium, nuclear matter, where m_i will be the momenta of particles in the chosen configuration.

The effects of the interactions which appear in the modified propagator change this single particle energy. In other words, the medium has become dispersive, and the energy relation is now given by

$$\epsilon_0(m_1) = \frac{m_1^2}{2M} + \sum_{i=2}^{A} [\langle m_1 m_i | t_0 | m_1 m_i \rangle - \text{exchange}]. \tag{4.2}$$

The significance of the subscript 0 on t will be explained later.

The modified single particle energy expression may now be used to replace the integral equation exhibited above by a set of coupled integral equations. The clue to this lies in the fact that the t-matrix which occurs in the propagators is to be evaluated while a different number of particles and holes are present in the field. (This propagation is said to take place 'off-the-energy-shell'.) We can actually classify the t-matrices according to the number of holes and particles present in the field while they are being evaluated.

An adequate notation is needed to express this situation. We start with all the particles in the chosen configuration occupying the states [30]

$$m_1, m_2, \ldots m_A \quad (A \to \infty \text{ for infinite systems}).$$

The energy of the particle in state m_1 is expressed by (2), where t_0 is the matrix evaluated with all particles in the chosen configuration. It is given by the integral equation

$$t_0 = v + vG_1t_0 \tag{4.3}$$

or in detail

$$\langle l_1l_2|t_0|m_1m_2\rangle = \langle l_1l_2|v|m_1m_2\rangle + \sum_{k_1k_2} \langle l_1l_2|v|k_1k_2\rangle\, G_1\, \langle k_1k_2|t_0|m_1m_2\rangle \tag{4.3'}$$

$$\frac{1}{G_1} = \epsilon_0(m_1) + \epsilon_0(m_2) - \epsilon_1\left(k_1;\, \begin{matrix} k_1 & k_2 \\ m_1 & m_2 \end{matrix}\right) - \epsilon_1\left(k_2;\, \begin{matrix} k_1 & k_2 \\ m_1 & m_2 \end{matrix}\right), \tag{4.4}$$

where $\epsilon_1(\ldots)$ are the single particle energies calculated when the particles occupying the states m_1 and m_2 of the chosen configuration have been excited to the states k_1 and k_2 (leaving holes in m_1 and m_2).

$$\epsilon_1\left(k_1;\, \begin{matrix} k_1 & k_2 \\ m_1 & m_2 \end{matrix}\right) = \frac{k_1^2}{2M} + \sum_{i=3} \left[\langle k_1m_i|t_1\left(k_1;\, \begin{matrix} k_1 & k_2 \\ m_1 & m_2 \end{matrix}\right)|k_1m_i\rangle - \text{exch}\right]. \tag{4.5}$$

where

$$t_1 = v + vG_2t_1 \tag{4.5'}$$

or in detail

$$\langle l_1l_2|\, t_1\left(k_1;\, \begin{matrix} k_1 & k_2 \\ m_1 & m_2 \end{matrix}\right) |k_1m_3\rangle = \langle l_1l_2|v|k_1m_3\rangle + \sum_{k_1{}^1,\, k_3} \langle l_1l_2|v|k_1^1k_3\rangle\, G_2\, \langle k_1^1k_3|t_1|k_1m_2\rangle \tag{4.6}$$

and

$$\frac{1}{G_2} = \epsilon_0(m_1) + \epsilon_0(m_2) + \epsilon_0(m_3) - \epsilon_1\left(k_1;\, \begin{matrix} k_1 & k_2 \\ m_1 & m_2 \end{matrix}\right) - \epsilon_2\left(k_1^1;\, \begin{matrix} k_1^1 & k_2 & k_3 \\ k_1 & k_2 & m_3 \\ m_1 & m_2 & m_3 \end{matrix}\right) - \epsilon_2\left(k_3;\, \begin{matrix} k_1^1 & k_2 & k_3 \\ k_1 & k_2 & m_3 \\ m_1 & m_2 & m_3 \end{matrix}\right). \tag{4.7}$$

[30] Convention of Ch. II, Sec. 4 is used for single particle indices.

The notation explains itself, e.g., the last term in (7) is the single particle energy in the configuration (k_1, k_2, k_3) obtained by letting m_1 and m_2 go into k_1 and k_2 (m_i, $i \geqq 3$ fixed) and then letting k_1 and m_3 go into k_1^1 and k_3 ($k_2 m_i$, $i \geqslant 4$ fixed) respectively.

Here we have a ladder which has the structure

$$\epsilon_0(m_i) \to t_0 \to G_1 \to \epsilon_1 \to t_1 \to G_2 \to \epsilon_2 \to t_2 \to G_3 \to \ldots.$$

All the equations have similar structure. The ith order single particle energy is expressed as

$$\epsilon_i(k_i; C_i) = \frac{k_i^2}{2M} + \sum_{j=i+2} [\langle k_i m_j | t(k_i, C_i) | k_i m_j \rangle - \text{exch.}] \tag{4.8}$$

where the configuration, C_i is given by

$$C_i = \begin{bmatrix} k_1^{i-1} & k_2 & k_3 & k_4 & \ldots & k_i & k_{i+1} \\ k_1^{i-2} & k_2 & k_3 & k_4 & \ldots & k_i & m_{i+1} \\ \vdots & \vdots & \vdots & \vdots & & \vdots & \vdots \\ \vdots & \vdots & \vdots & \vdots & & \vdots & \vdots \\ k_1^2 & k_2 & k_3 & k_4 & \ldots & m_i & m_{i+1} \\ k_1^1 & k_2 & k_3 & m_4 & \ldots & m_i & m_{i+1} \\ k_1 & k_2 & m_3 & m_4 & \ldots & m_i & m_{i+1} \\ m_1 & m_2 & m_3 & m_4 & \ldots & m_i & m_{i+1} \end{bmatrix}. \tag{4.9}$$

If the number of particles in the system is finite, $(= A)$, then the series will terminate at C_{A-1} since a new configuration does not result by letting the k's interact. They are arbitrary already. The t-matrix at the ith step is

$$\langle l_1 l_2 | t_i(k_i; C_i) | k_1^{i-1} m_{i+2} \rangle = \langle l_1 l_2 | v | k_1^{i-1} m_{i+2} \rangle$$
$$+ \sum_{k_1^i, k_{i+2}} \langle l_1 l_2 | v | k_1^i k_{i+2} \rangle \, G_{i+2} \, \langle k_1^i k_{i+2} | t_i(k_i, C_i) | k_i^{i-1} m_{i+2}) \rangle \tag{4.10}$$

and

$$\frac{1}{G_{i+1}} = \epsilon_0(m_1) + \sum_{j=2}^{i+1} [\epsilon_0(m_j) - \epsilon_{j-1}(k_{j-1}, C_{j-1})] - \epsilon_{i+1}(k_1^i, C_{i+1}). \tag{4.11}$$

One may also look upon the C_i as representing a diagram. Then C_{i-1} is a subdiagram of C_i and C_i itself is a subdiagram of C_{i+1}. The termination of this ladder of coupled integral equations occurs be-

$\epsilon_0(m_0) \to (m_1 m_2 | t_0 | m_1 m_2)$

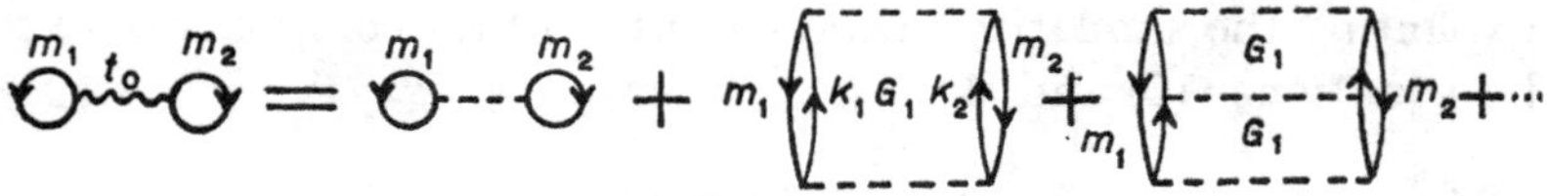

$G_1 \to \epsilon_1(k_1) \to (k_1 m_3 | t_1 | k_1 m_3)$

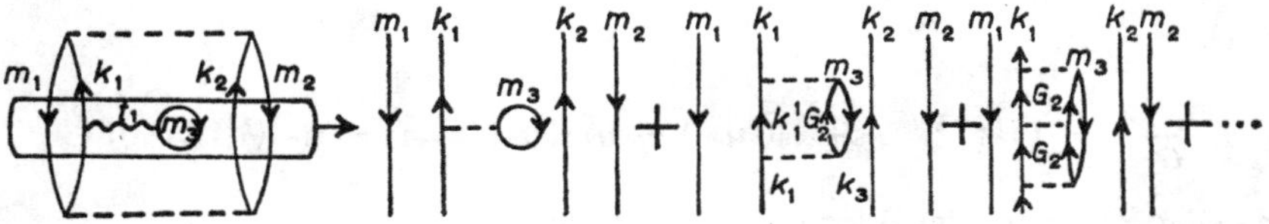

$G_2 \to \epsilon_2(k_1^1) \to (k_1^1 m_4 | t_2 | k_1^1 m_4)$

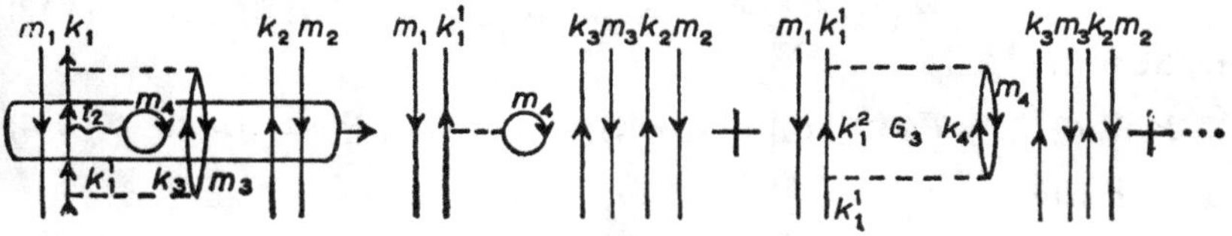

Also $\epsilon_3(k_3, \ldots) \to (k_3 m_4 | t_2 | k_3 m_4)$

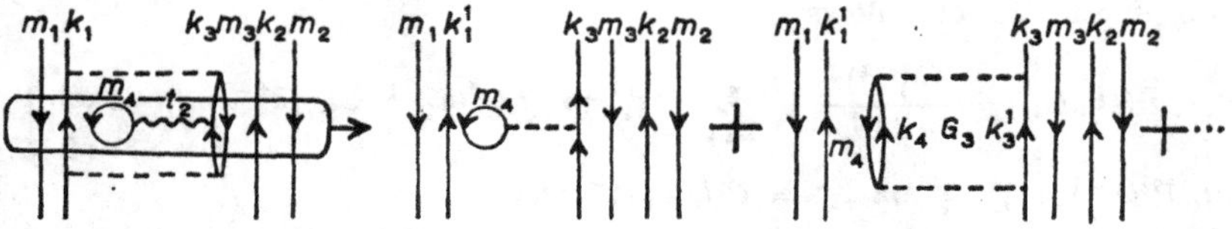

Fig. 28. Schematic representation of some of the diagrams summed by the ladder of integral equations of Brueckner and Gammel (B19). The last two lines of diagrams are equivalent. At each step we only show a part of the diagram.

cause if there are A particles in the chosen configuration, then there can be no diagrams which have more than A particles lines at a given time. This again is different from the case of field theory where there is no upper limit on the number of anti-particles. The excitation of a 'pair' in many body theory is thus different from the creation of a 'pair' in field theory. However, for an infinite medium the ladder of equations is infinite.

It was observed by Brueckner and Gammel (B19) that this infinite ladder of equations can be reduced to a single integral equation by exploiting the similarity in the structure of its components expressed by equations (10) and (11).

If one introduces a quantity, $\bar{P}$, defined by

$$\bar{P}(k_1^{i-1}, C_i) = \sum_{j=1}^{i} [\epsilon_0(m_j) - \epsilon_j(k_j, C_j)] \tag{4.12}$$

then

$$\frac{1}{G_{i+1}} = \bar{P}(k_1^{i-1}, C_i) + \epsilon_0(m_{i+1}) - \epsilon_{i+1}(k_{i+1}, C_{i+1}) - \epsilon_{i+1}(k_1^i, C_{i+1}),$$

or, using $\bar{P}$ also as a label for the configuration

$$\frac{1}{G_{i+1}} = \bar{P}(k_1^{i-1}, C_i) + \epsilon_0(m_{i+1}) - \epsilon_{i+1}(k_{i+1}; \bar{P}(k_{i+1}, C_{i+1})) - \epsilon_{i+1}(k_1^i; \bar{P}(k_1^i, C_{i+1})), \tag{4.13}$$

where, from (12),

$$\bar{P}(k_1^i, C_{i+1}) = \bar{P}(k_1^{i-1}, C_i) + \epsilon_0(m_{i+1}) - \epsilon_{i+1}(k_{i+1}; \bar{P}(k_{i+1}, C_{i+1})) \tag{4.14}$$

and analogously

$$\bar{P}(k_{i+1}, C_{i+1}) = \bar{P}(k_1^{i-1}, C_i) + \epsilon_0(m_{i+1}) - \epsilon_{i+1}(k_1^i, \bar{P}(k_1^i, C_{i+1})). \tag{4.15}$$

These equations might be considered as definitions of $\bar{P}(k_i, C_i)$. One may now rewrite (8) and (10) as follows

$$\epsilon_i[k_1^{i-1}, \bar{P}(k_1^i, C_i)] = \frac{(k_1^{i-1})^2}{2M} + \sum_{j=i+1} [\langle k_1^{i-1} m_j | t_i(\bar{P}(k_1^{i-1}, C_i)) | k_1^{i-1} m_j \rangle \text{—exch.}] \tag{4.16}$$

$$\langle l_1 l_2 | t_i[\bar{P}(k_1^{i-1}, C_i)] | k_1^{i-1} m_{i+1} \rangle = \langle l_1 l_2 | v | k_1^{i-1} m_{i+1} \rangle + \sum \langle l_1 l_2 | v | k_1^i k_{i+1} \rangle G_{i+1} \langle k_1^i k_{i+1} | t_i[\bar{P}(k_1^{i-1} C_i)] | k_1^{i-1} m_{i+1} \rangle, \tag{4.17}$$

where G_{i+1} is given in terms of $\bar{P}$ in (13).

Now $\bar{P}$ can be considered as a parameter independent of its arguments. That is, although it has definite values depending on its arguments one may drop the arguments and give it an infinite range. Then the set of equations to be solved becomes

$$\epsilon(l, \bar{P}) = \frac{l^2}{2M} + \sum_m [\langle lm | t(\bar{P}) | lm \rangle \text{—exch}] \tag{4.18}$$

$$\langle l_1 l_2 | t(\bar{P}) | lm \rangle = \langle l_1 l_2 | v | lm \rangle + \sum_{l_3 l_4} \langle l_1 l_2 | v | l_3 l_4 \rangle \, G \, \langle l_3 l_4 | t(\bar{P}) | lm \rangle \tag{4.19}$$

$$\frac{1}{G} = \bar{P} + \epsilon_0(m) - \epsilon(l_3, \bar{P}(l_3, l_4)) - \epsilon(l_4, \bar{P}(l_4, l_3)). \tag{4.20}$$

As in (14) and (15) one must solve the simultaneous equations

$$\bar{P}(l_3, l_4) = \bar{P} + \epsilon_0(m) - \epsilon(l_4, \bar{P}(l_4, l_3)) \tag{4.21a}$$

$$\bar{P}(l_4, l_3) = \bar{P} + \epsilon_0(m) - \epsilon(l_3, \bar{P}(l_3, l_4)). \tag{4.21b}$$

It can of course happen that (21) may not have any solutions or the solutions may not be unique. But the dependence of ϵ on $\bar{P}$ is only through the energy denominators in the t-matrices expansion and therefore it seems very likely that ϵ will be only a weakly varying monotonic function of $\bar{P}$ such that solutions of (21) will exist.

4a. *An approximation*

If one supposes that the propagation in all the excited states obeys approximately the same law, then one may break off the ladder at the second step. For calculating the energy of the particles in the chosen configuration one needs (3) and (4), then the propagator is written as

$$\frac{1}{G_e} = \epsilon_0(m_1) + \epsilon_0(m_2) - \epsilon_1(k_1, \bar{P}_e) - \epsilon_1(k_2, \bar{P}_e)$$
$$\equiv \bar{P}_e - \epsilon_1(k_1, \bar{P}_e) - \epsilon_1(k_2, \bar{P}_e). \tag{4.22}$$

The t-matrix which occurs in the derivation of $\epsilon_1(k, \bar{P}_e)$ depends on $\bar{P}_e$. Since the ladder is to be terminated here one puts in an approximate propagator at this point. The equations become

$$\epsilon_1(k_1, \bar{P}_e) = \frac{k_1^2}{2M} + \sum_{j=3} [\langle k_1 m_j | t(\bar{P}_e) | k_1 m_j \rangle - \text{exch.}] \tag{4.23}$$

$$\langle l_1 l_2 | t(\bar{P}_e) | k_1 m_j \rangle = \langle l_1 l_2 | v | k_1 m_j \rangle$$
$$+ \sum_{k_1{}^1, k_j} \langle l_1 l_2 | v | k_1^1 k_j \rangle \, G_e \, \langle k_1^1 k_j | t(\bar{P}_e) | k_1 m_j \rangle \tag{4.24}$$

$$\frac{1}{G_e} = \bar{P}_e - \Delta - \epsilon_1(k_1^1, \bar{P}_e) - \epsilon_1(k_j, \bar{P}_e). \tag{4.25}$$

The quantity $\bar{P}_e$ is arbitrary but needed only in the range from $2\epsilon_0(0)$ to $2\epsilon_0(p_F)$ corresponding respectively to the lowest and highest filled levels. The problem of solving the simultaneous equations (21) is now replaced by that of solving an extra integral equation. The parameter Δ takes account of the effect of other configurations; if the effect of propagation-off-the-energy-shell is weak one should have a fairly good approximation [31].

Brueckner and Gammel (B19) who worked in this approximation used a range of values for the parameter Δ from 0 to $\Delta_{\max} = \epsilon_0(p_F) - \epsilon_0(0)$ and found that the influence of this was rather weak thus justifying this approximation. They, however, believe that the numerical work involved in solving the exact equations (18)–(21) is not much more than in the present approximation.

5. THE LEVEL SHIFT AFTER PROPAGATOR MODIFICATION

Propagator modification is equivalent to choosing a suitable operator O in the Riesenfeld-Watson (R1) method. It is said that it takes into account the 'physical' dispersive characteristics of the medium. Any particle may be visualised as going through the medium making at each point a series of 'soft collisions' or 'forward scatterings' with other particles which do not change their state of motion. These 'forward scatterings' which occur through a t-matrix are in fact quite complicated, and are the cause of the dispersion in the medium. Finally there are also some 'hard collisions' in which a particle of the medium is excited above the Fermi sea.

The single particle energies defined above have no immediate connection with the level shift of the system as a whole. The total energy is not a simple sum of single particle energies. These single particle energies primarily were devices used to effect a more complete summation of higher order diagrams. In evaluating these one has to really calculate quantities like $\epsilon_i(k_i, C_i)$, $t(k_i, C_i)$, etc. which are many body quantities. Strictly speaking even the zero order $\epsilon_0(k_1)$ are not single particle quantities.

Returning to the t-matrix equation (2.5) for level shift and replacing the propagators and the t-matrices, one has

$$\begin{aligned}\Delta E = &\{\langle \textstyle\sum t_{\alpha 0} - U\rangle + \langle(\sum t_{\alpha 0} - U)\, G_1(\sum t_{\alpha 0} - U)\rangle + \dots\}_{\text{red}} \\ &- \{\langle(\textstyle\sum t_{\alpha 0} - U)\rangle\langle(\sum t_{\alpha 0} - U)\, G_1 G_1(\sum t_{\alpha 0} - U)\rangle + \dots\}_{\text{red}} \\ &- \{\textstyle\sum \langle t_{\alpha 0} G_1 t_{\alpha 0}\rangle + \dots\}_{\text{red}}.\end{aligned} \tag{5.1}$$

[31] However, this approximation was one of the reasons why Brueckner-Gammel (B19) missed the singularities of the t-matrix close to the Fermi surface (B33).

This might be misleading if one forgets the subscript **red**, which is to ensure that the number of terms is properly reduced in the above expression to allow for the fact that G_1 and $t_{\alpha 0}$ already imply summation of higher terms. The resulting series has not been investigated very well, because of the complexity of the summations involved.

It is easy to see, however, that already in Eq. (2.5) the second order terms disappear. In fact all terms with the same lines connected by successive t vertices disappear, and the first non-vanishing terms after first order are the linked terms [32] in the third order for the t-matrices. This is true for any type of propagator. It has been shown by numerical computation that the effect of these third order terms are small and one believes in fast convergence of the series (B15, K20).

Two questions still have to be answered. First: is the series really convergent? Second, and perhaps a more practical question: is it really worthwhile to have that formidable set of coupled integral equations? Could it not be that in this way one is summing up a whole lot of terms which are really not worth summing? The final expression (1) is no longer a power series in a small parameter. The integral equations definitely sum a vast number of terms of higher orders in the t-matrix but we also know definitely that we are omitting terms already in the third order. The question now is of the relative merit of neglecting these terms and including n^{th} order terms. It can be quite serious and it can also perhaps save us some labour.

The elimination of v in favour of t is necessary because otherwise we do not have a finite two-body matrix element but the subsequent propagator modification is a refinement whose usefulness is not quite so obvious.

The vertex and propagator modifications considered here by no means exhaust the possibilities of the rearrangements of perturbation expansion. For instance, a rearrangement very important for the electron gas has been carried out by Gell-Mann and Brueckner (B32, G10; see also K18, U1, W15 for other such approaches to the electron problem). However, for nuclear case, because of the nature of the inter-

[32] The so-called $1/A$-correction terms (B15) form a subclass of such terms. They can be quite important for light nuclei. Their influence on the properties of C^{12} has been discussed by Rodberg (R5).

action and the densities involved, this rearrangement is not important (B35).

6. CHOICE OF THE SINGLE PARTICLE POTENTIAL: SELF-CONSISTENCY

The aim of the theory is to provide a good single particle description of the many body system. With respect to the energies it means that the level shift (5.1) between the model energy, E_0, and the actual energy, $E_0 + \Delta E$ becomes as small as possible. Suppose one knows enough about (5.1) to trust it. Then the ΔE obtained from it can be further reduced by making an appropriate choice of the single particle potential U which is so far unspecified.

Apart from the criteria of simplicity U must be chosen to achieve the twin objectives of (a) reducing the level-shift and (b) making the level shift expansion rapidly convergent. Objective (b) can be realised if the choice of U is such as to induce cancellations in higher order terms. Since the structure of higher order terms is quite complex and not very well known, one can expect that this criterion will be only approximately fulfilled. One then falls back on making the level shift vanish in first order and thus satisfy (a). This leads to the use of a self-consistency procedure which has a long history in connection with solid state and atomic problems. In those problems the interaction v was sufficiently well behaved so that the t-matrix was not necessary. In nuclear problems the self-consistent single particle potential must be defined through the t-matrix. Of course there is no reason why in other cases the t-matrix should not be used. There also it is expected to improve the calculations. To make connection with older methods, they will be briefly discussed. Recall equation (2.4) for the level shift in terms of U and V

$$\Delta E = \left\{\langle \Sigma\, v_\alpha - U\rangle + \langle (\Sigma\, v_\alpha - U)\frac{1}{a}(\Sigma\, v_\alpha - U)\rangle + \ldots\right\}$$

$$-\left\{\langle \Sigma\, v_\alpha - U\rangle \, \langle (\Sigma\, v_\alpha - U)\frac{1}{a}\frac{1}{a}(\Sigma\, v_\alpha - U)\rangle + \ldots\right\}.$$

Note that its structure is still very similar to that of (5.1).

Since U is a single particle operator it can be expected to cancel only those elements of the two particle operator v_α in which interaction with a passive particle occurs, i.e., elements diagonal in one of the indices. Hartree's choice of U_i is (cf., equation (II. 3.19))

$$U_{\mathrm{H}} = \sum_m{}' \langle ml_1|v|ml_2\rangle\, \eta^*_{l_1}\eta_{l_2} \tag{6.1}$$

where the sum goes over all the occupied particles only. Hartree's choice is expressed in words as taking the single particle potential to be the sum of the interactions with all other particles in the medium. If exchange is also included then one has the Hartree-Fock choice. In earlier notation

$$U_{\mathrm{H-F}} = \sum_m{}' (ml_1, ml_2)\eta^*_{l_1}\eta_{l_2} \equiv \sum_m{}' \langle ml_1|v|ml_2 - l_2m\rangle\, \eta^*_{l_1}\eta_{l_2}. \tag{6.2}$$

With the choice (2) (or (1)) the level shift does not vanish in the first order. It becomes

$$\Delta E = \sum_{\text{pairs}} (m_1m_2, m_1m_2) - \sum_{m_1m_2} (m_1m_2, m_1m_2)$$

or

$$\Delta E = -\tfrac{1}{2}\langle U_{\mathrm{H-F}}\rangle. \tag{6.3}$$

The total energy now is $E = \langle T + \frac{1}{2}U_{\mathrm{H-F}}\rangle$ up to first order in $(v - u)$. The Hartree-Fock choice makes for cancellation of the diagram shown in Fig. 29a. The self-consistency procedure is schematically as follows:

$$(T_i, U_i)^0 \to \{\Phi_i\}^0, \quad \epsilon_0^0, \to U^1 \text{ by } (6.1, 6.2) \to \{\Phi_i\}^1,$$
$$\epsilon_0^1 = (T + \tfrac{1}{2}U^1) \to U^2 \text{ by } (6.1, 6.2) \to \text{etc.}, \quad \text{till } \epsilon_0^j = \epsilon_0^{j+1};$$
$$\{\Phi_i\}^j = \{\Phi_i\}^{j+1}$$

where the superscripts denote the number of iterations.

However, the definitions (1) and (2) of U are not quite so good from the point of view of the criterion (a) and (b) mentioned above because there is the residual term $\frac{1}{2}U$ already in the first order. The situation may be improved by making a different choice

$$U_{\mathrm{T}} = \sum_m{}' \langle ml_1|v(1 - \tfrac{1}{2}\delta_{l_1l_2})|ml_2 - l_2m\rangle\, \eta^*_{l_1}\eta_{l_2}. \tag{6.4}$$

This has the effect of making the level shift vanish in the first order. The energy consequently becomes, up to this order, $E = \langle T + U_{\mathrm{T}}\rangle$. As far as the self-consistency scheme is concerned it is not changed much because we have added only a diagonal operator to the earlier definition of U. Also, although the expression for the energy in terms of U has changed, its actual value has not. In view of the diagonal nature of the modification the wave functions are also unchanged. But it is an improvement over the Hartree-Fock choice in as much as there are more cancellations in the higher orders of the level shift expansion. This is the modified Hartree-Fock self-consistency condition proposed by Tobocman (T5) (Fig. 29b).

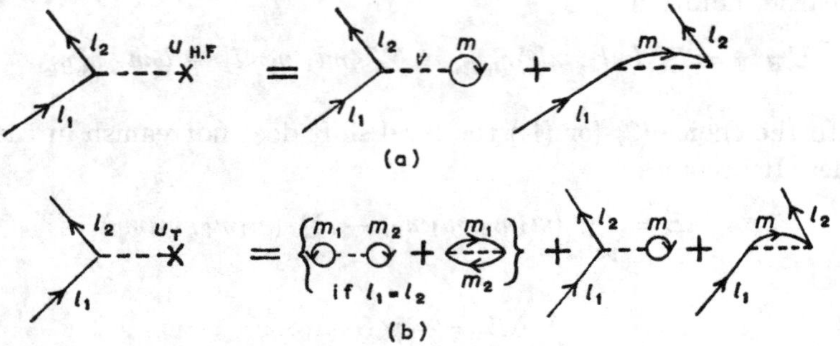

Fig. 29. Diagrammatic representation of the condition of self-consistency. (a) Hartree-Fock, (b) Tobocman.

The choice for the nuclear problem is naturally

$$U = \Sigma' \langle ml_1 | t[1 - \tfrac{1}{2}\delta_{l_1 l_2}] | ml_2 - l_2 m \rangle \, \eta^*_{l_1} \eta_{l_2} \tag{6.5}$$

for which one has to refer to the level shift expansion (5.1). The finite contributions to the level shift now first occur as linked (t-) diagrams in the third order. These have been shown to be small for the case of an infinite nuclear medium (B15, K20).

Brueckner, Bethe (e.g., B15) and others have used only the analogue of the Hartree definition for the single particle potential

$$U_{\mathrm{B}} = \sum_m{}' \langle ml_1 | t | ml_2 \rangle \, \eta^*_{l_1} \eta_{l_2}. \tag{6.6}$$

The choice

$$U_{\mathrm{E}} = \sum_m{}' \langle ml_1 | t | ml_2 - l_2 m \rangle \, \eta^*_{l_1} \eta_{l_2} \tag{6.7}$$

has been considered by Eden and Emery (E4), whereas Hugenholtz and Van Hove (H4) consider more elaborate choices which cancel second and higher order terms in t. These latter improvements have only third order effects on the energy and become important only for discussions of separation energies of last particles.

The occurrence of the t-matrix in (5) and (6) greatly complicates the problem because the t-matrix involves the wave functions of the Hamiltonian $T_i + U_i$. The scheme is now

$$(T_i + U_i^0) \to \{\Phi_i\}^0 \to t^0 \to \epsilon^1 \text{ and } U_i^1 \to \{\Phi_i\}^1 \to t^1 \to \epsilon^2,\ U_i^2 \to \text{etc.},$$

where the superscripts denote the number of iterations. Self-consistency is achieved when

$$\epsilon^i = \epsilon^{i+1}, \quad \{\Phi\}^i = \{\Phi\}^{i+1} \text{ and } t^i = t^{i+1}.$$

Simplification arises in the nuclear matter case because wave function self-consistency is not needed. The assumption that the system be uniform (i.e., invariant under translations) tells us that the wave functions will always be plane waves [33]. Further simplification occurs in the t-matrix equation, and will be discussed in more detail in Ch. IV.

This presentation of the self-consistency problem puts it in the right perspective and shows in what respect it differs from an exact treatment. The connection of the self-consistent procedure to a variational principle is discussed by Eden (E3, E4).

[33] The existence of stable, self-consistent solutions of this problem does not imply that the ground state of a large system of nucleons is uniform. Density, spin or isotopic spin waves in the medium are possible and may lead to states with energies lower than the uniform (Hartree-Fock) state. (See footnote 43.)

LIST OF IMPORTANT SYMBOLS IN CHAPTER III

(N.B. *Only new symbols or those with different meanings are listed here. Other symbols are as in Ch. II*)

$\dfrac{1}{a}$ free propagator

$A_{\mathrm{RS}}, A_t, A_\tau$ sum of all the terms having a particular structure in the R–S series with v-interaction (RS), t-interaction and τ-interaction vertices, (2.4, 2.5, 2.7)

$\dfrac{1}{b}$ propagator including the effect of 'forward scattering' by v-interaction

$B_{\mathrm{RS}}, B_t, B_\tau$ sum of all the terms of R–S series which cancel the contributions of unlinked clusters in A-type terms

C_t, C_τ terms of a certain (linked) structure which occur on eliminating v in favour of t- or τ-interaction

C_i configuration in which i-particles are excited out of the chosen configuration, (4.9)

$\dfrac{1}{e}$ propagator including the effects of 'forward scattering' by t-interaction

E total energy of the system. In this section and the next the perturbed energy is to be calculated in a self-consistent way so as to be equal to the unperturbed energy hence we use the same symbol for both – no confusion should arise, $(= \mathscr{E})$

g_α arbitrary two-particle propagator in a many body system

G, G_i generic symbol for propagators. G_i is the propagator for the configuration C_i. $i > 0$ corresponds to off-the-energy-shell propagation (4.11)

$\bar{P}(k, C_i)$ a quantity defined by means of single particle energy differences occurring in C_i and lower configurations. Also used as a label for the configuration C_i, (4.12)

$\bar{P}_e$ a parameter with continuous range used to approximate $\bar{P}(k, C_i)$ for all excited (intermediate) configurations

$t_{\alpha 0}$ t-operator for pair α in the ground state

t_{ii}	sum of the interactions of particle i with all the rest $(= \sum_{j \neq i} t_{ij})$
$t_i(k_i, C_i)$	t-operator for interactions of a particle in the state k_i and configuration C_i – off-the-energy-shell, $i > 0$, (4.10)
U	generic symbol for single particle potentials
V_{ii}	sum of interactions of particle i with all the rest $(= \sum_{j \neq i} V_{ij})$
α	subscript denoting a pair
δ_α	arbitrary constant occurring in the equation for t-matrix, (2.1)
Δ	a parameter introduced in the propagator to approximately take into account the effect of other configurations, (4.25)
$\epsilon(m_i)$	single particle energy
τ_α	an interaction operator, analogous to t-matrix, (2.6)

CHAPTER IV

METHODS OF SOLVING t-MATRIX EQUATIONS AND APPLICATION TO NUCLEAR PROBLEMS [34]

1. INTRODUCTION

1a. *Some characteristics of the nuclear problem*

In the foregoing we have developed the formal methods whose application to the nuclear problem will now be considered. We would first like to indicate briefly why these particular methods are being applied.

The data from experimental study of nuclei have fallen into several groups. Certain properties are associated with ground states: masses (or binding energies), density distributions (or shapes), moments and spins. Then there are similar properties for excited states and the probabilities of transition between various levels, and the data on alpha and beta decays. One tries to relate these data to as small a number of postulates as possible. In this endeavour special models are constructed to explain a limited number of phenomena. One of the most striking features of these models is that *nucleons seem to move nearly independently inside nuclei*. A great hope of nuclear physicists is to have a theory which in special cases will reduce to such models. However, since the basic things are the data, it is more logical to ask for a theory which will explain the measurements even if it does not quite reduce to some of the models.

The other important set data of nuclear physics comes from free nucleon-nucleon scattering. Here the cross sections and phase shifts indicate strong interactions. In particular, a strong repulsive core at a distance of about 0.4fm is indicated. This is in apparent contradiction with the evidence for near independent particle motion inside nuclei and thus gives rise to the most striking paradox of nuclear physics. The resolution of this paradox is sought by developing a theory which gives the properties of a system of many particles which

[34] Outside the field of nuclear physics such methods have been applied to the study of properties of liquid He^3 (B23) and positron annihilation in solids (K14). Use of the t-matrix has also been made in theory of Bosons (B32).

interact through *strong two-body interactions*. In particular, one is interested in seeing to what extent such a system may be represented as a collection of non-interacting (independent) particles.

It would be equally or even more satisfactory to ask for a connection at a different level. One might use meson theory to explain properties of nuclei and free nucleon interactions separately rather than try to derive the former from the latter. In view of the difficulties of the meson theory for even the simplest cases one is not encouraged to try to explain the properties of nuclei on the basis of meson theory. An attempt of this type was made by Johnson and Teller (J1). Their main result, based on a rather unrealistic meson theory, was that inside a nucleus the two-body force ceases to be strong – the nucleons do not interact directly but with a classical type meson field generated by the presence of all the nucleons. This implies absence of correlations between the nucleons and is in contradiction with the results of a number of experiments (B28) which show that there are substantial high momentum components in the nuclear wave function. As the simplest means of understanding these correlations we are led back to the point of view expressed at the end of the last paragraph.

Formal methods for treating systems having strong two-body interactions, in which the t-matrix has a central importance, have already been discussed together with the questions of defining single particle potentials. The main concern of this chapter is to discuss the techniques of solving the equations of this theory. From the outset it should be realised that most of these calculations can be carried out only under some approximations and the enormity of numerical work requires essential use of electronic computers. We shall discuss the physical reasoning behind various approximations and the details of calculations will be carried to a point beyond which standard numerical or other methods such as those used in earlier atomic or nuclear shell model calculations can be applied.

We shall pay most attention to the case of a system of infinite extension. Here an important simplification occurs if one can use the plane wave functions for the basic single particle motion. As a result a large class of matrix elements vanishes and there are other simplifications in the equations. In nature there are, of course, many systems of large extent to which such a theory can be applied directly, e.g., liquid He^3, liquid N^{15} and electrons in metals. In nuclear physics the theory is not directly applicable. One therefore speaks of 'nuclear matter', a hypothetical system of neutrons and protons in which there are no Coulomb interactions and which is of infinite extension. The theoretical results for such a system are compared with those obtained by extrapolating the experimentally observed trends in data connected with nuclei. Often these values are called 'experimental' which is rather misleading. All the important concepts needed for the study of nuclei already occur in the 'nuclear matter' problem and can be given a clear and a more complete discussion here in view of the mathematical simplifications mentioned above. Also the results obtained from this study serve as starting points for at least one of the methods applicable to nuclei.

It will be seen that schemes for calculating the properties of nuclei can be written down fairly easily after this. The main difficulty arises in actually carrying out the calculations. One has to proceed by way of a number of approximations whose validity is difficult to assess. Work on this aspect has barely begun and we give references to the relevant papers such as they are. Apart from this we have confined ourselves to pointing out the motivations, the so-called physical reasoning, behind the approximations. For the numerical results of the theory we refer the reader to the various original papers with a warning that the results have often been found to be sensitive to variations in parameters of the potentials and the type of numerical method used. In the last section we describe the nuclear shell model from the point of view of the present theory.

1b. *Standard forms: Bethe-Goldstone equation*

For purposes of solving the t-matrix equation it is convenient to write it in the form (B13)

$$\langle ij|t|rs\rangle = \langle ij|v|rs\rangle + \sum_{mn} \frac{\langle ij|v|mn\rangle\, \eta_n\eta_m\eta_m^*\eta_n^*\, \langle mn|t|rs\rangle}{\epsilon_r + \epsilon_s - \epsilon_m - \epsilon_n + \delta E}\,. \tag{1.1}$$

The choice of the propagator here is slightly different. Since

$$\eta_n\eta_m\eta_m^*\eta_n^* = (N_n - 1)(N_m - 1) \tag{1.2}$$

where N_n is the number of particles in the state n and can take on values 1 and 0 only, the sum extends only over the unoccupied states of the system, i.e. states above the Fermi surface. The single particle energies, ϵ_i, are given up to an arbitrary additive constant (vide infra) by

$$\epsilon_i = \frac{k_i^2}{2M} + \sum_j \langle ij|t(\delta E)|ij - ji\rangle. \tag{1.3}$$

The quantity δE is introduced approximately to take care of the propagation characteristics in excited states. $\delta E = 0$ for on-the-energy-shell-propagation and $\delta E \simeq \epsilon_F$ for off-the-energy-shell-propagation. The t-matrix is therefore written as $t(\delta E)$.

The modified propagator in (1) which restricts the summation to states above the Fermi sea enhances cancellations in the lower order terms of the t-expansions discussed in Ch. III. In higher orders the effect is in the opposite direction. One argues that the number of occupied states itself is small in the higher orders, hence this so-called exclusion principle effect of the propagator will be less important for higher orders. Moreover, the higher terms have to be small if the method is to work at all. Hence the net effect of the exclusion principle restriction in (1) would be to make the first terms of the level shift expansion a better representation of the whole series.

1b (i). THE TWO-PARTICLE WAVE-MATRIX

For actually calculating the t-matrix it is convenient to introduce a so-called Møller wave-matrix Ω_{12} by means of the relation

$$t_{12} = v_{12}\Omega_{12}. \tag{1.4}$$

From (1) and (4) one has

$$\langle l_4 l_3|\Omega_{12}|l_1 l_2\rangle = \langle l_4 l_3|l_1 l_2\rangle + \sum{}' \frac{\langle l_4 l_3|l_1' l_2'\rangle \langle l_1' l_2'|v_{12}\Omega_{12}|l_1 l_2\rangle}{\epsilon_{l_1} + \epsilon_{l_2} - \epsilon_{l_1'} - \epsilon_{l_2'} + \delta E} \tag{1.5}$$

where $\sum'$ represents the fact that the sum is restricted to unoccupied states only.

The physical significance of Ω_{12} becomes clear on referring to equation (I. 6.14). If there are only two particles in the system one has

$$M \equiv M_{12} = 1 + \frac{P_0}{e} t_{12}$$

$$t_{12} = v_{12} + v_{12} \frac{P_0}{e} t_{12} = v_{12} M_{12}$$

and the exact two-particle wave function

$$\Psi_{12} = M_{12} \Phi_{12} = \Omega_{12} \Phi_{12}. \tag{1.6}$$

Thus Ω_{12} is the matrix that takes the non-interacting two-particle wave function into the exact two-particle wave function.

1b (ii). First Brueckner approximation to the t-matrix

If the two particles are not in the nuclear medium then $\delta E = 0$ in the equation (5) for Ω_{12}, all the energies in the denominators are free particle energies and there is no exclusion principle restriction on the intermediate states of summation. In *that case* the matrix element $\langle \Phi_{12} | v | \Psi_{12} \rangle$ represents the scattering amplitude. But

$$\langle \Phi_{12} | v | \Psi_{12} \rangle = \langle \Phi_{12} | t | \Phi_{12} \rangle.$$

Hence one might approximately write for the nuclear matter case ($\hbar = 1$) (see however D4 and F7):

$$\langle \boldsymbol{k}_i, \boldsymbol{k}_j | t | \boldsymbol{k}_i, \boldsymbol{k}_j \rangle = -a_{\boldsymbol{k}_{ij}}(0) \frac{4\pi}{M\Omega}$$

$$\langle \boldsymbol{k}_i, \boldsymbol{k}_j | t | \boldsymbol{k}_j, \boldsymbol{k}_i \rangle = -a_{\boldsymbol{k}_{ij}}(\pi) \frac{4\pi}{M\Omega},$$

where $a_{\boldsymbol{k}_{ij}}(0)$ (or π) is the scattering amplitude in the forward (or backward) direction of two nucleons of relative momentum $\boldsymbol{k}_{ij} = \frac{1}{2}(\boldsymbol{k}_i - \boldsymbol{k}_j)$, and Ω in the numerical factor is the nuclear volume, which arises from integration over the total momentum (not to be confused with Ω_{12} which is a matrix). This approximation was used in the first papers of Brueckner and collaborators (B22, B27) on the binding energy problem and is also useful in nucleon-nucleus scattering problems at high energies (K15). Here a connection to the phase-shifts,

δ_l, may be made through the stationary state relation (B13)

$$a_k(0) = k^{-1} \sum_l (2l + 1) \tan \delta_l.$$

To obtain the single particle potential an average of the scattering amplitudes must be taken over the relative momentum distribution in the nucleus. A first approximation to the relative momentum distribution can be obtained by considering the nucleus as a Fermi gas.

The merit of this approximation is its simplicity and appeal to physical intuition. One says that for nuclear binding the most important effects are the coherent scatterings (forward and backward) of the nucleons in the nuclear medium.

One must now return to (1) and realise that the scattering in the nuclear medium must, above all, obey the exclusion principle and in this way the simple connection to the scattering amplitudes and the phase shifts is lost. Equation (5) still defines a two-particle wave function but it is now the two-particle wave function in the nuclear medium and is expected to differ from the free two-particles wave function.

1b (iii). Transformation to the coordinate space

The space wave function of two particles in states l_1 and l_2 is obtained from (5)

$$\langle \boldsymbol{r}_1 \boldsymbol{r}_2 | \Omega | l_1 l_2 \rangle = \langle \boldsymbol{r}_1 \boldsymbol{r}_2 | l_1 l_2 \rangle + \Sigma' \frac{\langle \boldsymbol{r}_1 \boldsymbol{r}_2 | l_1' l_2' \rangle \langle l_1' l_2' | v \Omega | l_1 l_2 \rangle}{\epsilon_{l_1} + \epsilon_{l_2} - \epsilon_{l_1'} - \epsilon_{l_2'} + \delta E} \tag{1.7}$$

or letting

$$\langle \boldsymbol{r}_1 \boldsymbol{r}_2 | \Omega | l_1 l_2 \rangle = \Psi_{l_1 l_2}(\boldsymbol{r}_1, \boldsymbol{r}_2) \tag{1.8a}$$

$$\langle \boldsymbol{r}_1 \boldsymbol{r}_2 | l_1 l_2 \rangle = \Phi_{l_1 l_2}(\boldsymbol{r}_1, \boldsymbol{r}_2) \tag{1.8b}$$

and

$$G(\boldsymbol{r}_1 \boldsymbol{r}_2, \boldsymbol{r}_1' \boldsymbol{r}_2') = \sum_{l_1' l_2'}{}' \frac{\Phi_{l_1' l_2'}(\boldsymbol{r}_1 \boldsymbol{r}_2) \Phi^*_{l_1' l_2'}(\boldsymbol{r}_1' \boldsymbol{r}_2')}{\epsilon_{l_1} + \epsilon_{l_2} - \epsilon_{l_1'} - \epsilon_{l_2'} + \delta E} \tag{1.8c}$$

$$\langle \boldsymbol{r}_1 \boldsymbol{r}_2 | v | \boldsymbol{r}_1' \boldsymbol{r}_2' \rangle = v(\boldsymbol{r}_1 \boldsymbol{r}_2, \boldsymbol{r}_1' \boldsymbol{r}_2') \tag{1.8d}$$

$$\Psi_{l_1 l_2}(\boldsymbol{r}_1 \boldsymbol{r}_2) = \Phi_{l_1 l_2}(\boldsymbol{r}_1 \boldsymbol{r}_2) + \int G(\boldsymbol{r}_1 \boldsymbol{r}_2, \boldsymbol{r}_1' \boldsymbol{r}_2') v(\boldsymbol{r}_1' \boldsymbol{r}_2', \boldsymbol{r}_1'' \boldsymbol{r}_2'') \cdot \Psi_{l_1 l_2}(\boldsymbol{r}_1'' \boldsymbol{r}_2'') \, \mathrm{d}\boldsymbol{r}_1' \, \mathrm{d}\boldsymbol{r}_2' \, \mathrm{d}\boldsymbol{r}_1'' \, \mathrm{d}\boldsymbol{r}_2''. \tag{1.9}$$

If v is a local potential, i.e. diagonal in $\boldsymbol{r}$-representation, and does not depend on the position of the centre of mass of the particles then

$$\langle \boldsymbol{r}_1\boldsymbol{r}_2|v\,|\boldsymbol{r}_1'\boldsymbol{r}_2'\rangle = \delta(\boldsymbol{r}_1+\boldsymbol{r}_2-\boldsymbol{r}_1'-\boldsymbol{r}_2')\delta(\boldsymbol{r}_1-\boldsymbol{r}_2-\boldsymbol{r}_1'+\boldsymbol{r}_2')v(|\boldsymbol{r}_1-\boldsymbol{r}_2|) \qquad (1.8e)$$

so that

$$\Psi_{l_1l_2}(\boldsymbol{r}_1\boldsymbol{r}_2) = \Phi_{l_1l_2}(\boldsymbol{r}_1\boldsymbol{r}_2) + \int G(\boldsymbol{r}_1\boldsymbol{r}_2, \boldsymbol{r}_1'\boldsymbol{r}_2')v(|\boldsymbol{r}_1'-\boldsymbol{r}_2'|)\Psi_{l_1l_2}(\boldsymbol{r}_1'\boldsymbol{r}_2')\,\mathrm{d}\boldsymbol{r}_1'\,\mathrm{d}\boldsymbol{r}_2'. \qquad (1.10)$$

A transformation to the centre of mass and relative coordinates $\boldsymbol{r} = \boldsymbol{r}_1 - \boldsymbol{r}_2$; $\boldsymbol{R} = \frac{1}{2}(\boldsymbol{r}_1 + \boldsymbol{r}_2)$ can be made only if the model wave functions occuring in the Green's function (8c) are separable in these coordinates. This seems to be possible in only two cases, viz., when the model wave functions are plane waves and when they are the wave functions in a harmonic oscillator potential. The first case corresponds to the so-called nuclear matter approximation which simulates the interior of heavy nuclei but is actually true only for a system of infinite extension. The second case has application to nuclei. Using harmonic oscillator wave functions, O^{16} has been discussed by Eden and collaborators (E4) and three and four body problem by Mang and Wild (M2). It will therefore be assumed that such a separation is possible.

If the quantum numbers for relative and centre of mass motion corresponding to l_1 and l_2, are respectively p, P, then (10) becomes

$$\Psi_{p,P}(\boldsymbol{r}, \boldsymbol{R}) = \Phi_{p,P}(\boldsymbol{r}, \boldsymbol{R}) + \int G(\boldsymbol{r}, \boldsymbol{R}, \boldsymbol{r}', \boldsymbol{R}')\, v(r')\Psi_{p,P}(\boldsymbol{r}'\boldsymbol{R}')\,\mathrm{d}\boldsymbol{r}'\,\mathrm{d}\boldsymbol{R}'. \qquad (1.11)$$

Because of separability $\Phi_{p,P}$ is actually a sum of the products of centre of mass and relative wave functions.

$$\Phi_{p,P}(\boldsymbol{r}, \boldsymbol{R}) \equiv \sum \Phi_p(\boldsymbol{r})\Phi_P(\boldsymbol{R}) \qquad (1.12)$$

and the Eq. (11) reduces to

$$\Psi_{p,P}(\boldsymbol{r}) = \Phi_p(\boldsymbol{r}) + \int G_P(\boldsymbol{r}, \boldsymbol{r}')v(r')\Psi_{p,P}(\boldsymbol{r}')\,\mathrm{d}\boldsymbol{r}' \qquad (1.13)$$

where

$$\Psi_{p,P}(\boldsymbol{r}) = \int \Phi_P^*(\boldsymbol{R})\Psi_{p,P}(\boldsymbol{r}\boldsymbol{R})\,\mathrm{d}\boldsymbol{R} \qquad (1.14)$$

and

$$G_P(\boldsymbol{r}, \boldsymbol{r}') = \sum_{p'(P)}{}' \frac{\Phi_{p'}(\boldsymbol{r})\Phi_{p'}^*(\boldsymbol{r}')}{\epsilon_{p,P} - \epsilon_{p',P} + \delta E_p}. \qquad (1.15)$$

$\sum'_{p'(P)}$ means that the sum over intermediate states p' of relative motion is to be carried out such that the exclusion principle requirement is obeyed with account taken of the fact that the centre of mass is in the state P. The energies $\epsilon_{p,P}$ are the energies in the new representation. It is desirable that there should be no coupling between the centre of mass and relative motion of the particles through the energies, i.e., it should be possible to write $\epsilon_{p,P} = \epsilon_p + \epsilon_P$ at least in some approximation. (Even then there would be a coupling through the restriction on the sum.) These conditions are fulfilled in the two cases mentioned above.

Applying the operator $(\epsilon_p - \mathscr{H})$ to Eq. (13), where $\mathscr{H}$ is the model two particle Hamiltonian in relative coordinates, one obtains

$$(\epsilon_p - \mathscr{H})\Psi_{p,P}(\boldsymbol{r}) = \int Q(\boldsymbol{r}, \boldsymbol{r}')v(r')\Psi_{p,P}(\boldsymbol{r}')\,\mathrm{d}\boldsymbol{r}', \tag{1.16}$$

where Q is a projection operator which takes into account the effect of the exclusion principle. Explicitly

$$Q(\boldsymbol{r}, \boldsymbol{r}') = \sum_{p'(P)}{}' \frac{\epsilon_p - \epsilon_{p'}}{\epsilon_{p,P} - \epsilon_{p',P} - \delta E_p} \Phi_{p'}(\boldsymbol{r})\Phi^*_{p'}(\boldsymbol{r}'). \tag{1.17}$$

It is clear that if the exclusion principle restriction on the summation is dropped, $\delta E_p = 0$ and $\epsilon_{p,P} = \epsilon_p + \epsilon_P$ then in (17) one has $Q(\boldsymbol{r}, \boldsymbol{r}') = \delta(\boldsymbol{r} - \boldsymbol{r}')$ and (16) reduces to the ordinary Schrödinger equation in relative coordinates with potential v. Thus once again the connection with the free particle case is made explicit.

Equations (16) and (13) are standard forms of the Bethe–Goldstone (B16) equation, usually abbreviated as the B–G Eq. In solving this equation there are two main sources of difficulty. The first and the more difficult is the effect of Pauli-principle, i.e. the effect of the Q operator. The second difficulty is connected with the existence of the hard core in the potential v. The latter makes it impossible to obtain a solution of (13) or (16) by iteration since a replacement of Ψ by Φ renders the integral infinite. Evaluation of these equations has been most extensively done in the nuclear matter or plane wave-approximation. This will be taken up next. Before that let us note that the matrix element of t in relative coordinates is given by

$$\langle p'|t(\delta E, P)|p\rangle = \sum_{p''} \langle p'|v|p''\rangle \langle p''|\Omega(\delta E, P)|p\rangle$$
$$= \int \mathrm{d}\boldsymbol{r}'\, \Phi_{p'}(\boldsymbol{r}')v(r')\Psi_{p,P}(\boldsymbol{r}'). \tag{1.18}$$

2. NUCLEAR MATTER CASE: EFFECTIVE-MASS APPROXIMATION

If the single particle potential inside the medium is expanded in powers of particle momentum k, one has [35]

$$U(k) = U_0 + U_1 k^2 + U_2 k^4 + \dots \tag{2.1}$$

At low enough value of momenta one can drop the higher terms. On assuming that for momenta occuring inside the medium it is sufficient to take only the first two terms, the energy becomes

$$\epsilon(k) = \frac{k^2}{2M^*} + U_0 \tag{2.2}$$

where

$$\frac{1}{2M^*} = \frac{1}{2M}(1 + 2MU_1) \tag{2.3}$$

M^* is the effective mass. The quantities U_0, U_1 and M^* are in general functions of position of the particle. This is a very useful approximation and takes into account the main features of the momentum dependence of the potential [36].

In the uniform nuclear matter problem both U_0 and U_1 are constants independent of position and the problem becomes particularly simple. The energy denominators are now separable. Since the quantum numbers are now three-dimensional momenta, we have

$$\boldsymbol{l}_1' + \boldsymbol{l}_2' = \boldsymbol{l}_1 + \boldsymbol{l}_2 = \boldsymbol{P}, \quad \boldsymbol{l}_1 - \boldsymbol{l}_2 = 2\boldsymbol{p}, \quad \boldsymbol{l}_1' - \boldsymbol{l}_2' = 2\boldsymbol{p}'.$$

$$\epsilon(\boldsymbol{l}_1) + \epsilon(\boldsymbol{l}_2) - \epsilon(\boldsymbol{l}_1') - \epsilon(\boldsymbol{l}_2') = \frac{1}{2M^*}[l_1^2 + l_2^2 - l_1'^2 - l_2'^2]$$

$$= \frac{1}{M^*}(p^2 - p'^2). \tag{2.4}$$

[35] It was realised quite early in the history of nuclear physics that the presence of exchange forces between the particles, and also the exchange part of the expectation value of ordinary forces will give rise to a velocity-dependent single particle potential (v.V1, B38). The concept of effective mass is implicit in this assumption and the corresponding expressions already occur, e.g., in reference (L8).

[36] Limitations of this approximation have been pointed out among others by Moszkowski and Scott (M7) and by Levinger *et al.* (L4).

Further, if there is no complication due to hard cores, it is convenient to work with the t-matrix equation (1.1) directly. With plane wave functions

$$\langle \boldsymbol{l}_1'\boldsymbol{l}_2'|v|\boldsymbol{l}_1\boldsymbol{l}_2\rangle = w(\boldsymbol{p}, \boldsymbol{p}')(2\pi)^3\delta(\boldsymbol{l}_1' + \boldsymbol{l}_2' - \boldsymbol{l}_1 - \boldsymbol{l}_2) \tag{2.5}$$

where $w(\boldsymbol{p}, \boldsymbol{p}')$ is the Fourier transform of the two-particle potential,

$$w(\boldsymbol{p}, \boldsymbol{p}') = \int v\,(\boldsymbol{r}_{12}, \boldsymbol{r}_{12}')\, e^{i\{\boldsymbol{p}\cdot\boldsymbol{r}_{12} - \boldsymbol{p}'\cdot\boldsymbol{r}_{12}'\}}\, d\boldsymbol{r}_{12}\, d\boldsymbol{r}_{12}'. \tag{2.6}$$

If the potential is local, $v(\boldsymbol{r}_{12}, \boldsymbol{r}_{12}') = v(r_{12})\delta(\boldsymbol{r}_{12} - \boldsymbol{r}_{12}')$, and we have $w(\boldsymbol{p}, \boldsymbol{p}') = w(|\boldsymbol{p} - \boldsymbol{p}'|)$ which is the usual form of the Fourier transform. The factor $(2\pi)^3$ in (5) is introduced for correct normalisation of the matrix element. In the case of box normalisation for a single discrete state $\boldsymbol{l}_1'$, fixed by giving the other three states $\boldsymbol{l}_2'\boldsymbol{l}_1\boldsymbol{l}_2$, the matrix element w is given by (6). In order to obtain the same result in the continuum case where one must integrate over the $\boldsymbol{l}_1'$ momentum space with a volume element $(2\pi)^{-3}\, d\boldsymbol{l}_1'$, the integrand should be given by (5) (see footnote 30 in ref. B15).
Similarly

$$\langle \boldsymbol{l}_3\boldsymbol{l}_4|t|\boldsymbol{l}_1\boldsymbol{l}_2\rangle = \langle \boldsymbol{p}'|t|\boldsymbol{p}\rangle\,(2\pi)^3\delta(\boldsymbol{P} - \boldsymbol{P}')$$

and from (1.1)

$$\langle \boldsymbol{p}'|t|\boldsymbol{p}\rangle = w(\boldsymbol{p}', \boldsymbol{p}) + (2\pi)^{-3}M^* \int' d\boldsymbol{p}''w(\boldsymbol{p}', \boldsymbol{p}'')\,\frac{1}{p^2 - p''^2}\,\langle \boldsymbol{p}''|t|\boldsymbol{p}\rangle, \tag{2.7}$$

where the limits on the integral are determined by the exclusion principle. In the special case $\boldsymbol{P} = 0$, $\boldsymbol{l}_1' = -\boldsymbol{l}_2' = \boldsymbol{p}''$ and the integration over $\boldsymbol{p}''$ extends only outside a sphere of radius k_F.

In this approximation it is possible to solve this integral equation exactly if the two-particle potential is separable (Y2, M4) in momentum space. (In which case it will not be local in the coordinate space, i.e. it will have no factor $\delta(\boldsymbol{r}_{12} - \boldsymbol{r}_{12}')$ cf., (6).) The general form of a separable potential is

$$w(\boldsymbol{p}, \boldsymbol{p}') = \sum_i \tilde{w}_i(\boldsymbol{p})\tilde{w}_i(\boldsymbol{p}').$$

In the particular case of only one member in the sum

$$w(\boldsymbol{p}', \boldsymbol{p}) = \tilde{w}(\boldsymbol{p}')\tilde{w}(\boldsymbol{p}) \tag{2.8}$$

and the exact solution of (7) is given by

$$\langle \boldsymbol{p}'|t|\boldsymbol{p}\rangle = w(\boldsymbol{p}',\boldsymbol{p})\left[1 - \frac{M^*}{(2\pi)^3}\int \mathrm{d}\boldsymbol{p}''\,\frac{\{\tilde{w}(\boldsymbol{p}'')\}^2}{p''^2 - p^2}\right]^{-1}. \tag{2.9}$$

Separable potentials for nucleon-nucleon scattering have been studied by Yamaguchi and Yamaguchi and more recently by Mitra (Y2, M4). They have been applied to the present problem by Nigam and Sundaresan and others (N3, M4).

This approximation with the use of separable potentials in view of the simple and exact solution (9) is particularly suited for investigating various qualitative features of the t-matrix.

2a. *Singularities of the t-matrix*

One of the results obtained most easily from (9) is that the t-matrix is singular for values of p such that

$$1 = \frac{M^*}{(2\pi)^3}\int \mathrm{d}\boldsymbol{p}''\,\frac{\{\tilde{w}(\boldsymbol{p}'')\}^2}{p''^2 - p^2}. \tag{2.10}$$

There exists a considerable literature on the subject (C2, E5, M3 and others). It has not been shown that the singularities are due to any peculiarities of the effective mass approximation or the use of separable potential. As a matter of fact the singularities are believed to exist for all effectively attractive potentials. On integrating such a t-matrix to obtain energy one gets a singular result. Inclusion of more complicated terms in the t-matrix equation, such as those due to hole-hole interaction does not alter the situation. Only the details of the location of the singularities are changed. These singularities of the t-matrix always occur for values of single particle momentum p immediately below or above the Fermi momentum p_{F}. The occurrence of the singularities is independent of the interaction strength – the weaker the interaction the closer the momenta have to be to the Fermi surface and more nearly equal to zero has their total momentum to be. In the calculations of Brueckner and collaborators, the most elaborate of which is that by Brueckner and Gammel (B19), these singularities were 'missed' primarily because of the wide momentum mesh used in the computations. The calculated momenta closest to the Fermi surface were still too far away to give noticeably high contributions. Fortunately, this is perhaps not a very serious difficulty. The divergencies may be avoided by a suitable redefinition of the t-matrix following a procedure used in the treatment of cases where the chosen configuration is degenerate (B15 and Ch. IV, sec. 7b). This essentially consists in dropping those intermediate states in the t-matrix equation which led to the divergence. The contribution of the 'dangerous' intermediate states may then be evaluated using the methods of the theory of superconductivity (B33, C6). Even if these contributions are small, as believed by Brueckner (B33, B35), we are here faced with an apparent breakdown of perturbation theory. We shall return to these questions in Ch. V, sec. 5.

3. NUCLEAR MATTER CASE: PARTIAL WAVE EXPANSION

In view of the connection of Eq. (1.16) to the free particle scattering case, further refinements within the framework of the plane wave approximation (infinite medium, nuclear matter problem) use the partial wave analysis (B19, K21). This treatment is very similar to that of the free particle scattering problem.

For the sake of completeness, spin will be included from now on. Let $\chi_s^{m_s}$ be the spin function of the two nucleons where s can take on values 0 and 1 for singlet and triplet states respectively. The model wave function for relative motion is then given by

$$\Phi_{\boldsymbol{k}}(\boldsymbol{r}) = \exp(\mathrm{i}\boldsymbol{k}\cdot\boldsymbol{r})\chi_s^{m_s} = \sum_{l=0}^{\infty} C_l j_l(kr) Y_l^0(\hat{\boldsymbol{k}}\hat{\boldsymbol{r}})\chi_s^{m_s}, \tag{3.1}$$

where

$$C_l = (2l+1)\mathrm{i}^l \left(\frac{4\pi}{2l+1}\right)^{\frac{1}{2}}$$

$\hat{\boldsymbol{x}}$ unit vector in the direction $\boldsymbol{x}$

$\int \mathrm{d}\hat{\boldsymbol{x}}$ integration over angles of $\boldsymbol{x}$

$f(\hat{\boldsymbol{x}}_1, \hat{\boldsymbol{x}}_2)$ a function of angles θ_1, φ_1 and θ_2, φ_2 between the vectors $\boldsymbol{x}_1$ and $\boldsymbol{x}_2$ and the arbitrary reference frame.

In presence of the tensor force (A1, B20) l is not a good quantum number but the total J is, and hence (1) should be expanded in the eigenfunctions of J (B21).

$$|(ls)JM\rangle \equiv \mathscr{Y}_{Jls}^{M} = \sum_{m_s} (l, m_J - m_s, s, m_s | J m_J) Y_l^{m_J - m_s} \chi_s^{m_s} \tag{3.2}$$

or inverting the relation

$$Y_l^{m_l}\chi_s^{m_s} = \sum_{J=l-1}^{J=l+1} (lm_l\, s\, m_s|JM)\mathscr{Y}_{Jls}^M. \tag{3.3}$$

Hence from (1)

$$\Phi_{\boldsymbol{k}}(\boldsymbol{r}) = \sum_{J=0}^{\infty}\sum_{l=J-1}^{J+1} C_l j_l(kr)(l\,0\, sm_s|Jm_s)\mathscr{Y}_{Jls}^{m_s}(\hat{\boldsymbol{k}}, \hat{\boldsymbol{r}}), \tag{3.4}$$

$\mathscr{Y}_{Jls}^M$ are such that for $s = 0$ only $J = l$ is allowed and for $s = 1$ only $J = l,\ l \pm 1$ are allowed. In ref. (B17) these functions have been denoted by $F_l^{Jm_Js}$.

The tensor force $S_{12} = (3r^{-2}(\boldsymbol{\sigma}_1\cdot\boldsymbol{r})(\boldsymbol{\sigma}_2\cdot\boldsymbol{r}) - (\boldsymbol{\sigma}_1\cdot\boldsymbol{\sigma}_2))$, conserves J, s and parity but not the orbital angular momentum l, hence for $s = 1$ and arbitrary J only the states with $l = J \pm 1$ will be coupled. Because of parity conservation the state $s = 1$, $l = J$ is left undisturbed. Therefore, in decomposing Ψ one must provide for distinguishing the states $l = J \pm 1$ generated by the unperturbed (orthogonal) functions with $l = J + 1$ and $l = J - 1$. This is accomplished by putting two subscripts on the perturbed partial wave function, $\mathscr{U}$, according to the following conventions:

$$j_{J\mp1}(kr)\mathscr{Y}^M_{J,J-1,s}(\hat{\boldsymbol{k}}, \hat{\boldsymbol{r}}) \to \mathscr{U}^{Js}_{J\mp1,J-1}(kr)\mathscr{Y}^M_{J,J-1,s}(\hat{\boldsymbol{k}}, \hat{\boldsymbol{r}}) + \mathscr{U}^{Js}_{J\mp1,J+1}(kr)\mathscr{Y}^M_{J,J+1,s}(\hat{\boldsymbol{k}}, \hat{\boldsymbol{r}}) \quad \text{for } s = 1$$

$$j_J(kr)\mathscr{Y}^M_{J,J,s}(\hat{\boldsymbol{k}}, \hat{\boldsymbol{r}}) \to \mathscr{U}^{Js}_{J,J}(kr)\mathscr{Y}^M_{J,J,s}(\hat{\boldsymbol{k}}, \hat{\boldsymbol{r}}) \quad \text{for } s = 0,\ 1.$$

With the above convention the decomposition of the perturbed wave function becomes

$$\Psi_{\boldsymbol{k}}(\boldsymbol{r}) = \sum_{J=0}^{\infty}\sum_{l=J-1}^{J+1} C_l \sum_{l'=J-1}^{J+1} \mathscr{U}^{Js}_{ll'}(r)(l\,0\, sm_s|Jm_s)\mathscr{Y}^{m_s}_{Jl's}, \tag{3.5}$$

where the index P on Ψ has been dropped. An angular momentum expansion of the Green's function is also needed,

$$G(\boldsymbol{r}, \boldsymbol{r}') = \sum_{l=0}^{\infty} C_l(-\mathrm{i})^l G_l(r, r')Y_l^0(\hat{\boldsymbol{r}}, \hat{\boldsymbol{r}}'). \tag{3.6}$$

To obtain the explicit form of $G_l(r, r')$ some approximations have to be made which will be discussed later.

Using these expansions in (1.13) and dropping the sum over J, since J is a constant of motion, one has

$$\sum_{ll'} C_l \mathscr{U}^{Js}_{ll'}(r)(l\,0\,sm_s|Jm_s)\mathscr{Y}^{m_s}_{Jl's} = \sum_l C_l j_l(kr)(l\,0\,sm_s|Jm_s)\mathscr{Y}^{m_s}_{Jls}$$
$$+ \sum_l \int \mathrm{d}\boldsymbol{r}' C_l(-\mathrm{i})^l G_l(r, r') Y^0_l(\hat{\boldsymbol{r}}, \hat{\boldsymbol{r}}')v(r')$$
$$\cdot \sum_{l'l''} C_l \mathscr{U}^{Js}_{l'l''}(r')\mathscr{Y}^{m_s}_{Jl''s}(\hat{\boldsymbol{k}}, \hat{\boldsymbol{r}}')(l'\,0\,sm_s|Jm_s). \tag{3.7}$$

Integration over the angles of $\boldsymbol{r}'$ can be performed to give

$$(2l+1)\left(\frac{4\pi}{2l+1}\right)^{\frac{1}{2}} \int \mathrm{d}\hat{\boldsymbol{r}}' Y^0_l(\hat{\boldsymbol{r}}, \hat{\boldsymbol{r}}')v(r')\mathscr{Y}^{m_s}_{Jl''s}(\hat{\boldsymbol{k}}, \hat{\boldsymbol{r}}') = 4\pi \mathscr{Y}^{m_s}_{Jls}(\hat{\boldsymbol{k}}, \hat{\boldsymbol{r}})v^{Js}_{ll''}(r'), \tag{3.8}$$

where use has been made of the relations

$$Y^0_l(\hat{\boldsymbol{r}}, \hat{\boldsymbol{r}}') = \left(\frac{4\pi}{2l+1}\right)^{\frac{1}{2}} \sum_{m=-l}^{l} Y^{m*}_l(\hat{\boldsymbol{k}}, \hat{\boldsymbol{r}}')Y^m_l(\hat{\boldsymbol{k}}, \hat{\boldsymbol{r}})$$

and

$$v^{Js}_{ll''}(r) = \int \mathrm{d}\hat{\boldsymbol{r}}\, \mathscr{Y}^{m_s*}_{Jls}(\hat{\boldsymbol{k}}, \hat{\boldsymbol{r}})v(r)\mathscr{Y}^{m_s}_{Jl''s}(\boldsymbol{k}, \hat{\boldsymbol{r}}). \tag{3.9}$$

Actually the last quantity is fairly simple. One can write

$$v^{Js}_{ll''}(r) = S^{Js}_{ll''}v_{\mathrm{T}}(r) \quad \text{for tensor forces,}$$
$$= \delta_{ll''}v_{\mathrm{C}}(r) \quad \text{for central forces.}$$

The numbers $S^{Js}_{ll'}$ are, for example, listed by Ashkin and Wu (A1). $v_{\mathrm{T}}(r)$ and $v_{\mathrm{C}}(r)$ are the radial dependences of the tensor and central forces respectively.

Finally, one can remove the reference to the angle variables from (7) altogether by taking scalar product w.r.t. to $\mathscr{Y}^{m_s}_{JLs}(\hat{\boldsymbol{k}}, \hat{\boldsymbol{r}})$ and using (9), of course at the same time changing the dummy index l to l' on the left hand side and integrating over $\hat{\boldsymbol{r}}$.

$$\sum_{l'} C_{l'} \mathscr{U}^{Js}_{l'L}(l'\,0\,sm_s|Jm_s) = C_L j_L(kr)(L\,0\,sm_s|Jm_s)$$
$$+ \sum_{l'l''} \int r'^2\,\mathrm{d}r'\, 4\pi G_L(r, r')v^{Js}_{Ll''}(r')C_{l'}\mathscr{U}^{Js}_{l'l''}(r')(l'\,0\,sm|Jm_s).$$

Then multiplying throughout by $(L'\,0\,sm_s|Jm_s)$ and summing over m_s and using the orthogonality of the Clebsch-Gordan coefficients

$$\sum_{m_s=-1}^{1} (L'\,0\,sm_s|Jm_s)(l'\,0\,sm_s|Jm_s) = \delta_{L'l'}$$

one obtains

$$\mathscr{U}^{Js}_{ll'}(r) = \delta_{ll'}j_l(kr) + 4\pi \sum_{l''} \int r'^2\,\mathrm{d}r'\, G_{l'}(r, r')v^{Js}_{ll''}(r')\mathscr{U}^{Js}_{ll''}(r'). \tag{3.10}$$

These are the equations which have to be solved and the t-matrix is given in terms of these by means of the relation (1.18).

3a. *Diagonal elements of t-matrix*

For calculating the total and single particle energies the diagonal elements are the most important ones. We have

$$\langle \boldsymbol{k}, s, m_s|t|\boldsymbol{k}, s, m_s\rangle = \sum_J \int \mathrm{d}\boldsymbol{r} \{ \sum_{l=J-1}^{J+1} C_l j_l(kr) \mathscr{Y}^{m_s}_{Jls}(\hat{\boldsymbol{k}}, \hat{\boldsymbol{r}})$$

$$\times (l\,0\,sm_s|Jm_s)\} v(r) \{ \sum_{l'=J-1}^{J+1} C_{l'} \sum_{l''=J-1}^{J+1} \mathscr{U}^{Js}_{l'l''}(r) \mathscr{Y}^{m_s}_{Jl''s}(\hat{\boldsymbol{k}}, \hat{\boldsymbol{r}})(l'\,0\,sm_s|Jm_s). \quad (3.11)$$

Using (9) and summing over the magnetic quantum numbers (which gives $\delta_{ll'}$) one has

$$\sum_{m_s=-1}^{+1} \langle \boldsymbol{k}sm_s|t|\boldsymbol{k}sm_s\rangle$$

$$= 4\pi \sum_J \sum_{l=J-1}^{J+1} (2l+1) \int r^2 \,\mathrm{d}r\, j_l(kr) \sum_{l'=J-1}^{J+1} v^{Js}_{ll'}(r) \mathscr{U}^{Js}_{ll'}(r). \quad (3.12)$$

For singlet states $s = 0$, and only $l = l' = J$ is possible. Hence

$$\langle \boldsymbol{k}\,0\,0|t|\boldsymbol{k}\,0\,0\rangle = 4\pi \sum_J (2J+1) \int_0^\infty r^2 \,\mathrm{d}r\, j_J(kr) v_{JJ}(r) \mathscr{U}_{JJ}(r). \quad (3.13)$$

For triplet states the sum can be written as a trace of a matrix product.

$$\sum_{m_s=-1}^{1} \langle \boldsymbol{k}\,1\,m_s|t|k\,1\,m_s\rangle = 4\pi \sum_J (2J+1) \int_0^\infty r^2 \,\mathrm{d}r \,\mathrm{Tr}\, \{jv\mathscr{U}\}, \quad (3.14)$$

where the matrix $\{jv\mathscr{U}\}$ is given by

$$\{jv\mathscr{U}\}$$

$$= \begin{bmatrix} j_{J-1} & 0 & 0 \\ 0 & j_J & 0 \\ 0 & 0 & j_{J+1} \end{bmatrix} \begin{bmatrix} v_{J-1,J-1} & 0 & v_{J-1,J+1} \\ 0 & v_{JJ} & 0 \\ v_{J+1,J-1} & 0 & v_{J+1,J+1} \end{bmatrix} \begin{bmatrix} \mathscr{U}_{J-1,J-1} & 0 & \mathscr{U}_{J-1,J+1} \\ 0 & \mathscr{U}_{JJ} & 0 \\ \mathscr{U}_{J+1,J-1} & 0 & \mathscr{U}_{J+1,J+1} \end{bmatrix}.$$

The superscripts J and s have been suppressed as in Eq. (13).

The above expressions are just the sums over spin magnetic quantum numbers under the influence of a possible tensor force. The details of the spin dependence of the t-matrix can be handled in this way. Further the interaction between the two particles also depends on the isobaric spin state, T. If, however, the Coulomb forces are neglected then the forces do not depend on T_Z (charge symmetry). The single particle potential experienced by the particle is then written as

$$U_0(\boldsymbol{k}_i, \eta_i, \sigma_i)$$
$$= \sum_{\boldsymbol{k}_j \eta_j \sigma_j} [\langle \boldsymbol{k}_i, \eta_i, \sigma_i; \boldsymbol{k}_j, \eta_j, \sigma_j | t | \boldsymbol{k}_i, \eta_i, \sigma_i; \boldsymbol{k}_j, \eta_j, \sigma_j \rangle - \text{exchange}], \quad (3.15)$$

where η_i are isobaric spin and σ_i are (ordinary) spin coordinates. Since the wave functions obey the generalised Pauli principle (space exchange $\times$ spin exchange $\times$ isobaric spin exchange $= -1$) the exchange terms are the same as the direct terms and a factor of two is obtained. Equation (15) can be resolved in spin and isobaric spin singlet and triplet states by using the relations

$$\alpha_i \alpha_j = \chi_1^1, \quad \alpha_i \beta_j = 2^{-\frac{1}{2}}(\chi_1^0 + \chi_0^0),$$
$$\beta_i \beta_j = \chi_1^{-1}, \quad \beta_i \alpha_j = 2^{-\frac{1}{2}}(\chi_1^0 - \chi_0^0) \quad (3.16)$$

$\sigma_i = \alpha_i$ or β_i corresponding to spin up and spin down respectively; with similar relations for isobaric spin wave functions. Since σ_i and η_i are fixed there are three states each for isobaric and mechanical spin of the other particles. Two of these bring a factor of $\frac{1}{2}$ each. Combined with the factors from the exchange the net factor is just $\frac{1}{2}$. The sum over T_Z quantum numbers in view of the assumed charge symmetry (no Coulomb forces) gives the factor $(2T+1)$ and finally

$$U_0(\boldsymbol{k}_i(\eta_i, \sigma_i))$$
$$= \sum_{\boldsymbol{k}_j} \tfrac{1}{2} \sum_{T=0}^{1} (2T+1) \sum_{s=0}^{1} \sum_{m_s=-1}^{+1} \langle \boldsymbol{k}_i \boldsymbol{k}_j | t(s, m_s, T) | \boldsymbol{k}_i \boldsymbol{k}_j \rangle. \quad (3.17).$$

Implicit in the derivation of (17) is the fact that t is independent of the sign of m_s which occurs in it only through some Clebsch-Gordan coefficients (see Eq. (11)).

It is conceivable that the potential depends upon both the charge [37] and the spin state. In that case one will also have to consider four Fermi momenta corresponding to the four distinct Fermi gases consisting of protons with spin up, protons with spin down, neutrons up

[37] The difference in neutron and proton potentials in an actual nucleus is only vaguely connected with this effect in as much as the dominant effect there comes from the Coulomb potential.

and neutrons down. This complication is avoided if it is assumed that (17) shows no dependence [38] on η_i and σ_i. This approximation is good enough for present purposes in view of the charge symmetry and infinite medium assumptions. It will require closer study when Coulomb effects are to be included. In this approximation the single particle energy is

$$\epsilon_0(k_i) = \frac{k_i^2}{2m} + U_0(k_i) \tag{3.18}$$

and the average binding energy per particle is given by

$$E_{\mathrm{Av}} = \frac{\int_0^{k_F} k^2 \, \mathrm{d}k [k^2/2M + \frac{1}{2}U_0(k)]}{\int_0^{k_F} k^2 \, \mathrm{d}k} . \tag{3.19}$$

The factor of half in the potential energy comes from the fact that each pair has to be counted only once in finding the total energy. (See, however, discussion in Ch. III, sec. 5.)

To complete the discussion we need a method of evaluating the angular momentum expansion of the Green's function and the contribution of the hard core. These we take up in turn.

3b. *Approximations in Green's function expressions*

The quantity of interest is from [39] equation (1.15)

$$G_{\mathbf{P}}(\boldsymbol{r}, \boldsymbol{r}') = \sum_{p'} \frac{\exp [\mathrm{i}\boldsymbol{p}' \cdot (\boldsymbol{r} - \boldsymbol{r}')]}{\epsilon_{p,P} - \epsilon_{p',P} + \delta E} f(P, p'), \tag{3.20}$$

where the function $f(P, p') = 0$ for states forbidden by Pauli principle.

It was seen that no difficulty arises if we have $\boldsymbol{P} = 0$ and make the effective mass approximation. The Green's function then depends only on the angle between $\boldsymbol{p}'$ and $(\boldsymbol{r} - \boldsymbol{r}')$ so that an expansion like

[38] This is supposed to be the case for closed shell nuclei, and is one of the reasons why the first applications of the method have been to close shell nuclei. (The other reason is that for closed shell nuclei certain sums over quantum numbers assume simple forms.)

[39] The notation explains itself, p and k both are commonly used for (relative and single-particle) momenta.

(6) was obtainable in a straightforward way. As these approximations are removed, one faces an additional coupling between the directions of $\boldsymbol{p}'$ and $\boldsymbol{P}$ through both $f(P, p')$ and the energy denominators.

Brueckner and Gammel (B19) have removed this coupling by replacing the angle dependent quantities both in the numerator and the denominator by their averages over the angles of P. The manner of performing these averages is of some interest and is given below. Similar approximations are considered in references (M7) and (S8). A more elaborate treatment of the case of non-zero centre of mass motion has been given by Werner (W10).

3b (i). Pauli principle

The two intermediate momenta in question are $\boldsymbol{p}' \pm \frac{1}{2}\boldsymbol{P}$ and the exclusion principle requires that

$$\begin{aligned} f(P, p') &= 0 \text{ for } |\boldsymbol{p}' \pm \tfrac{1}{2}\boldsymbol{P}| < p_F \\ &= 1 \text{ for } |\boldsymbol{p}' \pm \tfrac{1}{2}\boldsymbol{P}| > p_F. \end{aligned} \tag{3.21}$$

The approximation to this is as follows (B13)

$$\sum_{\boldsymbol{p}'} f(P, p') \to \int \mathrm{d}\hat{\boldsymbol{p}}' f'(P, p')$$

where

$$\begin{aligned} f'(P, p') &= 0 \text{ if } (p'^2 + \tfrac{1}{4}P^2)^{\frac{1}{2}} < p_F \\ &= 1 \text{ if } |\boldsymbol{p}' - \tfrac{1}{2}\boldsymbol{P}| > p_F \\ &= p'^2 + \tfrac{1}{4}P^2 - p_F^2/p'P \text{ otherwise.} \end{aligned} \tag{3.22}$$

To begin with the exclusion effect is small, as the success of the first Brueckner approximation (B22) shows, and therefore the present refinement is supposed to take account of the exclusion principle to a good enough degree.

3b (ii). The energy denominators

If we take terms higher than quadratic in momenta in single particle energies then we have, with $\boldsymbol{p}_\pm = \boldsymbol{p}' \pm \frac{1}{2}\boldsymbol{P}$, from Eq. (2.1) and (18)

$$\begin{aligned} \epsilon_1(p_+) + \epsilon_1(p_-) &= 2U_0 + U_1(p_+^2 + p_-^2) + U_2(p_+^4 + p_-^4) + \dots \\ &\cong 2U_0 + 2U_1(p'^2 + \tfrac{1}{4}P^2) + 2U_2\left[p'^4 + (\boldsymbol{p}'\cdot\boldsymbol{P})^2 + \frac{P^4}{16} + \frac{p'^2P^2}{2}\right]. \end{aligned}$$

The approximation consists in replacing

$$p_\pm^2 \to \tfrac{1}{4}P^2 + p'^2 \pm \frac{1}{\sqrt{3}} f'^{\frac{1}{2}}(P, p')pP \tag{3.23}$$

f' is defined in (22). The replacement is exact for $p > P$, up to quartic terms. Appeal is then made to the success of effective mass approximation and the fact that quartic terms perhaps become important only for $p > P$.

With these approximations the Green's function expansion can be made and is given by

$$G_l(r, r') = 4\pi \int_0^\infty \frac{p'^2\, dp'\, j_l(p'r) j_l(p'r') f'(P, p')}{\epsilon_{p,P} - \epsilon_{p',P} + \delta E}, \tag{3.24}$$

where an approximate form of energy denominators is used.

3c. *Treatment of the hard core potential* (B16, B19)

The only easily accessible method for solving the integral equation (1.13) or the corresponding t-matrix equation, is a partial wave expansion of the type discussed which gives for $\mathcal{U}(r)$ an equation of the form

$$\mathcal{U}(r) = \mathcal{U}_0(r) + 4\pi \int_0^\infty G(r, r') v(r') \mathcal{U}(r') r'^2\, dr'. \tag{3.25}$$

These equations have to be solved by iteration. However, if the potential $v(r')$ has a hard core, such as is now well established in the nucleon-nucleon potential, the iteration procedure breaks down because of divergence in the integral.

A method of overcoming this difficulty has been given by Bethe and Goldstone (B16). Here we follow the presentation of Brueckner and Gammel (B13, B19). The integral in (25) is broken up into two parts.

$$\int_0^\infty = \int_0^{r_c} + \int_{r_c}^\infty, \tag{3.25'}$$

where r_c is the core radius. The second part is well behaved because there are no infinities there. In the first part one makes the replacement

$$v(r)\mathcal{U}(r) = \lambda \delta(r - r_c), \quad r \leqslant r_c, \tag{3.26}$$

where λ is a constant to be determined by the condition that the wave function $\mathcal{U}(r)$ vanishes at the boundary. With this replacement (25) gives

$$\mathcal{U}(r) = \mathcal{U}_0(r) + 4\pi r_c^2 \lambda G(r, r_c) + 4\pi \int_{r_c}^\infty G(r, r') v(r') \mathcal{U}(r') r'^2\, dr'. \tag{3.27}$$

Requiring $\mathcal{U}(r_c) = 0$, we have

$$\lambda = -[\mathcal{U}_0(r_c) + 4\pi \int_{r_c}^{\infty} G(r_c, r')v(r')\mathcal{U}(r')r'^2 \, dr']/4\pi r_c^2 G(r_c, r_c). \tag{3.28}$$

Substituting this back in (27) we get

$$\mathcal{U}(r) = S(r) + 4\pi \int_{r_c}^{\infty} F(r, r')v(r')\mathcal{U}(r')r'^2 \, dr' \tag{3.29a}$$

with

$$S(r) = \mathcal{U}_0(r) - \mathcal{U}_0(r_c) \frac{G(r, r_c)}{G(r_c, r_c)} \tag{3.29b}$$

$$F(r, r') = G(r, r') - G(r, r_c)G(r_c, r')/G(r_c, r_c). \tag{3.29c}$$

This is directly applicable to the singlet state equation. For the triplet case we have coupled equations (10) but a similar treatment goes through. The replacements needed in (10) are

$$j_l(kr) \to S_l(kr) = j_l(kr) - j_l(kr_c) \frac{G_l(r, r_c)}{G_l(r_c, r_c)} \tag{3.30a}$$

$$G_l(r, r') \to F_l(r, r') = G_l(r, r') - G_l(r, r_c) \frac{G_l(r_c, r')}{G_l(r_c, r_c)} . \tag{3.30b}$$

It is clear, however, that the equations obtained in this way are not equivalent to the actual equations (10) or (25). The wave functions obtained from analogues of (29a) do not satisfy the original B–G equation (1.16). The error made in this substitution has been considered for the case of a hard core potential, without outside attraction. It has been found that for pure two body scattering the replacement leads to an exact result (B13) whereas for the nuclear matter the corrections although not entirely vanishing are small (B16, B19). One then offers the following reason for expecting reasonable results in the general case: For a well behaved purely attractive potential the use of Born approximation often gives a good result (L4, S16, also see Fig. 30). Hence one may use iteration procedures outside the core radius and take core effects into account by the above replacement.

Undoubtedly, this point deserves further investigation since it forms the basis of most numerical results of this theory. Kowalski and

Feldman (K21) have treated similar equations in greater mathematical detail.

3d. *Qualitative effects of the hard core potentials*

These effects were first studied by Bethe and Goldstone (B16). Gomes, Weisskopf and Walecka (G4) in a more detailed paper on the subject have further clarified the 'physical' processes occurring. In later works improvements have been made by considering more realistic potentials and removing some other approximations. Here we just give a few simple pictures which emerged from the earlier studies, and which have not been substantially altered by later work.

For non-interacting particles in a relative S-state with zero total momentum the wave function is

$$\Psi = \frac{\mathscr{U}(r)}{r} = \lim_{k\to 0} \frac{\sin kr}{kr} = 1.$$

This is shown in Fig. 30a, the wave function $\mathscr{U}(r)$ passes through the origin where it is zero, as it must be. Introduction of an attractive potential, say a square well pulls the wave function into the well. It still has to pass through the origin and at large distances merge into the original wave function, Fig. 30b. Since the wave function does not deviate much from the free particle value, the iteration procedure should be applicable. In particular, the Born approximation ($\mathscr{U} = r$) should be fairly good but the values so obtained for the t-matrix and energies would be slightly too small.

If there is a hard core then the wave function must vanish at $r \leqslant r_c$, (Fig. 30c). Asymptotically the original wave function is still good. Breakdown of the Born approximation is evident. The actual wave function is always less than its asymptotic value.

A case closer to the physical situation is a hard core at $r = r_c$ followed by an attractive well up to $r = r_b$ (Fig. 30d). Curve 3 in Fig. 30d is the actual solution. For some values of r it is higher than the asymptotic wave function shown as curve 1. It is always higher than the wave function for a hard core without attraction (curve 2). At the hard core it must go down to zero. The interesting thing that emerges here is that the Born approximation iteration using non-interacting wave function outside the core radius should be even

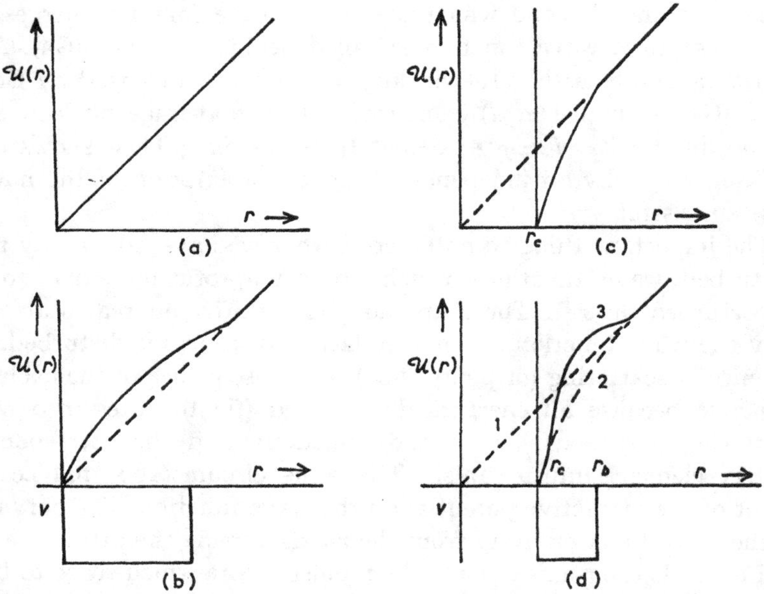

Fig. 30. Zero momentum S-state wave function for (a) free particles, (b) particles with a pure attraction between them, (c) particles with a repulsive core and (d) particles with a mixed potential.

better than in the case of pure attraction (Fig. 30b). For $r \leqslant r_c$ the breakdown is of course evident. This provides the motivation for the separation of contributions in Eq. (25′) and for the procedures adopted in some other works (E4, L4, M7) where the potential is broken up into two parts, one containing the hard core and the other purely attractive, in order that one may be treated by ordinary perturbation theory and the t-matrix methods may be used only for the other. If appropriately done, this may even improve convergence [40].

[40] This possibility of splitting the potential introduces another degree of arbitrariness into the theory and definitely makes the situation more complex. Actually one should aim at generating some small quantity in powers of which the level shift may be consistently expanded. Any splitting of the potential usually gives an expansion in two quantities only one of which is small. This may often obscure the real issue. Similar objections may be raised with greater justification to the works where the contributions to level shift from different 'physical' effects are separated. The word 'physical', like 'qualitative', is often misleading in such contexts.

The distance beyond which the actual wave function merges into the asymptotic wave function is called healing distance (G4). A remarkable result is that the healing distance as calculated by Gomes *et al.* (G4) is small ($\simeq \frac{2}{3}d$) compared with the average nucleon separation in nuclei $d \simeq \rho^{-\frac{1}{3}} \simeq (\frac{3}{2}\pi^2)^{\frac{1}{3}}\, \lambda\!\!\!^{-}_{\mathrm{F}} \simeq 1.66$ fm. This shows in a striking way why an independent particle description of the nucleus is so successful.

The important thing to note here is that we have taken only those perturbed wave functions which are asymptotically equal to the unperturbed ones [41]. There are no phase shifts, no real scattering, only the wave function at small (relative) distances is disturbed. The absence of scattering (or phase shift) is a consequence of the exclusion principle because all energetically allowed (final) states into which scattering may occur are already filled up and the corresponding matrix elements must vanish. The same circumstance reduces the effect of the attractive potential on the wave function. The curvature of the wave function away from the r-axis outside the attractive well and towards the r-axis outside the repulsive core which serve to bring the wave function back to the unperturbed form, are thus due to Pauli principle.

It is clear that the effects on the wave functions depend on the parameters of the potential. In the nuclear problem the hard core radius ($\simeq$ 0.4 fm) is small compared to the range of nuclear forces ($\simeq$ 2.3 fm). The strength of the attractive potential is such as to cause the average separation between nucleons to be much larger ($\simeq$ 1.7 fm) than the core radius. In this sense one talks of nuclear forces being weak and long ranged and the nucleus being a low density gas (compared to the density at close packing of the hard cores). The success of independent particle models depends on these circumstances.

The circumstances in the other infinite fermion systems, e.g., liquid

[41] This choice of boundary conditions is needed only in order that the perturbed and unperturbed spectra have the same structure – this being the usual requirement for the validity of the perturbation theory (see Ch. I, sec. 1). Other types of solutions of the B–G equation, the so-called localised solutions, are known to exist (L3 and other references). In a very specialised model such solutions led to lower energy than the present ones (K9). Full implications of the existence of these localised solutions have not yet been understood. It is believed that they are connected with the existence of superfluid type of states for the system (see Ch. V, sec. 5).

He^3, are known to be different (B23). Here the attractive part of the potential outside the hard core is very strong and the core radius is comparable to the range of the potential. There are corresponding differences in the properties of the system.

3e. *Scheme for energy calculations* (*B19*)

The iteration procedure is started off by using the energy expressions

$$\epsilon_0(k_0) = \frac{k_0^2}{2M}, \qquad \epsilon_1(k_i) = \frac{k_i^2}{2M}.$$

The energy denominators made this way are used to calculate the Green's function $G_l(r, r')$ given by (24) which involves approximate treatment of Pauli principle. Quantities $G_l(r_c, r)$, $G_l(r_c, r_c)$ etc. are then used to obtain the functions $S_l(kr)$ and $F_l(kr)$ defined in (30a, b). These are then used in Eq. (29a) for $\mathscr{U}$'s to take effects of the hard core into account. The $\mathscr{U}$'s are obtained by iteration and t-matrix elements given by (11) are calculated. Single particle potentials are obtained from these according to (17). The single particle energies are obtained according to (18) by taking terms up to the fourth power in momentum; approximation for decoupling of the orientations of $\boldsymbol{P}$ and $\boldsymbol{p}$ is made according to (23). The cycle is now repeated until self-consistent values of t and E are obtained. There is no wave function self-consistency required since these are always plane waves.

A schematic representation of the procedure is as follows:

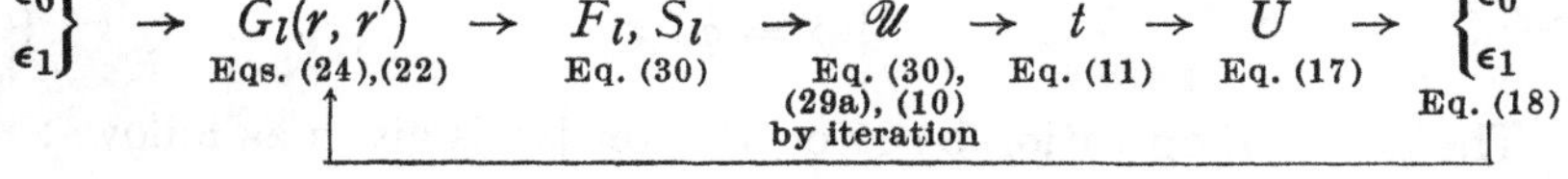

$$\begin{array}{ccccccccccccc}
 & & & & & & v(r) & & & & & & \\
 & & & & & & \downarrow & & & & & & \\
\left.\begin{matrix}\epsilon_0\\ \epsilon_1\end{matrix}\right\} & \rightarrow & G_l(r, r') & \rightarrow & F_l, S_l & \rightarrow & \mathscr{U} & \rightarrow & t & \rightarrow & U & \rightarrow & \left\{\begin{matrix}\epsilon_0\\ \epsilon_1\end{matrix}\right. \\
 & & \text{Eqs. (24),(22)} & & \text{Eq. (30)} & & \text{Eq. (30), (29a), (10) by iteration} & & \text{Eq. (11)} & & \text{Eq. (17)} & & \text{Eq. (18)}
\end{array}$$

(with a return arrow from Eq. (18) back to $G_l(r, r')$)

The simple appearance of the above scheme is deceptive. Any accurate calculation is a formidable task negotiable only by high speed computers. Even so, in evaluating the above expressions the sum over $Jll'l''$ has to be restricted from practical considerations. From a knowledge of the potential one can argue which $v_{ll'}^{Js}$ are important. A discussion of the nucleon-nucleon potential (B19, G5) shows that there is practically no contribution to nuclear binding from odd-parity singlet or triplet states. Therefore, in most works a potential which is non-zero only in even-parity states is considered. Further it is found that terms with $l \geqslant 4$ have little effect. Thus only $l = 0, 2$ terms need be considered. The couplings through $\mathscr{U}$ equations to higher (J and l) states are neglected. The only states considered in (B19) were $({}^3S_1 + {}^3D_1)$, ${}^3D_2$, ${}^3D_3$, ${}^1S_0$ and ${}^1D_1$.

4. THE *t*-MATRIX AND EFFECTIVE SINGLE PARTICLE POTENTIALS IN COORDINATE SPACE (B17)

Nuclei actually occurring in nature have finite extensions in coordinate space. This must be reflected in the *t*-matrix and single particle potentials appropriate to these nuclei. Also such aspects of the single particle potential as the spin-orbit coupling etc. are best discussed in coordinate representation. In the work described in the previous sections, the momentum or the energy representation of these quantities was obtained. A transformation to coordinate representation is therefore needed. Both the operators in question are expected to be non-local in $\boldsymbol{r}$. To fix the notation let us first consider some formal manipulations with non-local operators.

Let $|\alpha\rangle$ be the abstract ket for the state and A a non-local operator. The wave function in the $\boldsymbol{r}$- and $\boldsymbol{p}$-representation are then denoted respectively by $\langle \boldsymbol{r}|\alpha\rangle$ and $\langle \boldsymbol{p}|\alpha\rangle$. For instance in the plane wave case α is a momentum state, say $\alpha \equiv \boldsymbol{k}$, then explicitly

$$\langle \boldsymbol{r}|\alpha \equiv \boldsymbol{k}\rangle = \frac{1}{\sqrt{\Omega}} e^{i\boldsymbol{k}\cdot\boldsymbol{r}}, \langle \boldsymbol{p}|\alpha \equiv \boldsymbol{k}\rangle = \delta(\boldsymbol{p} - \boldsymbol{k}). \tag{4.1}$$

In general let

$$\langle \boldsymbol{r}|\alpha\rangle = \varphi_\alpha(\boldsymbol{r}). \tag{4.2}$$

The effect of operation by A on the ket $|\alpha\rangle$ is given as follows:

$$A|\alpha\rangle = \int A|\boldsymbol{r}'\rangle \langle \boldsymbol{r}'|\alpha\rangle \, d\boldsymbol{r}' = \int A|\boldsymbol{r}'\rangle \varphi_\alpha(\boldsymbol{r}') \, d\boldsymbol{r}' \tag{4.3}$$

and

$$\langle \boldsymbol{r}|A|\alpha\rangle = \int \langle \boldsymbol{r}|A|\boldsymbol{r}'\rangle \varphi_\alpha(\boldsymbol{r}') \, d\boldsymbol{r}' \tag{4.4}$$

and

$$\langle \beta|A|\alpha\rangle = \int \langle \beta|\boldsymbol{r}'\rangle \langle \boldsymbol{r}'|A|\boldsymbol{r}\rangle \langle \boldsymbol{r}|\alpha\rangle \, d\boldsymbol{r} \, d\boldsymbol{r}' \tag{4.5}$$

or

$$\langle \beta|A|\alpha\rangle = \int \varphi_\beta^*(\boldsymbol{r}') \langle \boldsymbol{r}'|A|\boldsymbol{r}\rangle \varphi_\alpha(\boldsymbol{r}) \, d\boldsymbol{r} \, d\boldsymbol{r}'. \tag{4.6}$$

There exists the inverse of the relation (6)

$$\langle \boldsymbol{r}' | A | \boldsymbol{r} \rangle = \sum_{\alpha,\beta} \varphi_\beta(\boldsymbol{r}') \langle \beta | A | \alpha \rangle \varphi^*_\alpha(\boldsymbol{r}). \tag{4.7}$$

4a. *The t-matrix*

The t-matrix is a two particle operator, hence

$$\langle \boldsymbol{r}_1' \boldsymbol{r}_2' | t | \boldsymbol{r}_1 \boldsymbol{r}_2 \rangle = \sum_{\alpha\beta\gamma\delta} \varphi_\alpha(\boldsymbol{r}_1') \varphi_\beta(\boldsymbol{r}_2') \langle \alpha\beta | t | \gamma\delta \rangle \varphi^*_\gamma(\boldsymbol{r}_1) \varphi^*_\delta(\boldsymbol{r}_2). \tag{4.8}$$

In momentum representation the general matrix element of t between states of relative momentum $\boldsymbol{p}$ and $\boldsymbol{p}'$ and total momentum $\boldsymbol{P}$ of the particles (spins are implicit) is given by

$$\langle \boldsymbol{p}_1' \boldsymbol{p}_2' | t | \boldsymbol{p}_1 \boldsymbol{p}_2 \rangle = \delta(\boldsymbol{P} - \boldsymbol{P}') \langle \boldsymbol{p}' | t(\boldsymbol{P}) | \boldsymbol{p} \rangle = \delta(\boldsymbol{P} - \boldsymbol{P}') (\Phi_{\boldsymbol{p}'}, v \Psi_{\boldsymbol{p},\boldsymbol{P}}). \tag{4.9}$$

Suppressing the index $\boldsymbol{P}$ on which the dependence is weak in any case, using this relation in (8), and assuming v to be a non-local operator

$$\langle \boldsymbol{r}_1' \boldsymbol{r}_2' | t | \boldsymbol{r}_1 \boldsymbol{r}_2 \rangle$$

$$= \delta(\boldsymbol{R} - \boldsymbol{R}') \sum_{\boldsymbol{p}',\boldsymbol{p}} \int \Phi_{\boldsymbol{p}'}(\boldsymbol{r}') \Phi^*_{\boldsymbol{p}'}(\boldsymbol{r}'') \langle \boldsymbol{r}'' | v | \boldsymbol{r}''' \rangle \Psi_{\boldsymbol{p}}(\boldsymbol{r}''') \Phi^*_{\boldsymbol{p}}(\boldsymbol{r}) \, \mathrm{d}\boldsymbol{r}'' \, \mathrm{d}\boldsymbol{r}'''. \tag{4.10}$$

Using the relations

$$\sum_{\boldsymbol{p}} \Phi_{\boldsymbol{p}}(\boldsymbol{r}) \Phi^*_{\boldsymbol{p}}(\boldsymbol{r}') = \delta(\boldsymbol{r} - \boldsymbol{r}') \tag{4.11}$$

and

$$\sum_{\boldsymbol{p}} \to (2\pi)^{-3} \int \mathrm{d}\boldsymbol{p} \tag{4.12}$$

we have

$$\langle \boldsymbol{r}_1' \boldsymbol{r}_2' | t | \boldsymbol{r}_1 \boldsymbol{r}_2 \rangle = \frac{\delta(\boldsymbol{R} - \boldsymbol{R}')}{(2\pi)^3} \int \mathrm{d}\boldsymbol{p} \, \mathrm{d}\boldsymbol{r}'' \, \Phi^*_{\boldsymbol{p}}(\boldsymbol{r}) \langle \boldsymbol{r}' | v | \boldsymbol{r}'' \rangle \Psi_{\boldsymbol{p}}(\boldsymbol{r}''). \tag{4.13}$$

Alternatively, (13) may be derived by noting that $t = v\Omega$ and using (1.8a) etc. If v in Eq. (13) is a local operator, (all except the Yamaguchi potentials are of this type (Y2, M4)):

$$\langle \boldsymbol{r}' | v | \boldsymbol{r} \rangle = v(r) \delta(\boldsymbol{r} - \boldsymbol{r}')$$

then we have

$$\langle \boldsymbol{r}_1' \boldsymbol{r}_2' | t | \boldsymbol{r}_1 \boldsymbol{r}_2 \rangle = \frac{\delta(\boldsymbol{R} - \boldsymbol{R}')}{(2\pi)^3} \int \mathrm{d}\boldsymbol{p} \, \Phi^*_{\boldsymbol{p}}(\boldsymbol{r}) v(r') \Psi_{\boldsymbol{p}}(\boldsymbol{r}'). \tag{4.14}$$

In practical computations it is necessary to separate the spin contributions. Therefore writing these indices explicitly we have in relative coordinates only

$$\langle \boldsymbol{r}'s'm'|t|\boldsymbol{r}sm\rangle = \frac{1}{(2\pi)^3}\int \mathrm{d}\boldsymbol{k}\, \Phi_{\boldsymbol{k}s'm'}(\boldsymbol{r})v(\boldsymbol{r}')\Psi_{\boldsymbol{k}sm}(\boldsymbol{r}'). \qquad (4.15)$$

Using the angular momentum expansions of Φ and Ψ given in (3.4) and (3.5) together with (3.9) and the relations

$$\int \mathrm{d}\Omega_k Y_l^0(\hat{\boldsymbol{k}}\hat{\boldsymbol{r}})v(r)\mathscr{Y}^m_{Jl's}(\hat{\boldsymbol{k}}\hat{\boldsymbol{r}}) = \sum_{l''}\left(\frac{2\pi}{2l+1}\right)^{\frac{1}{2}} v^{Js}_{ll''}(r)\mathscr{Y}^m_{Jl''s}(\hat{\boldsymbol{r}}, \hat{\boldsymbol{r}}')\delta_{ll'} \qquad (4.16)$$

and

$$\chi_s^{m'*}\mathscr{Y}^m_{Jls}(\hat{\boldsymbol{r}}, \hat{\boldsymbol{r}}') = \sum (lm'sm - m'|Jm)Y_l^{m-m'}(\hat{\boldsymbol{r}}, \hat{\boldsymbol{r}}') \qquad (4.17)$$

one gets

$$\langle \boldsymbol{r}'s'm'|t|\boldsymbol{r}sm\rangle = \frac{4\pi}{(2\pi)^3}\int k^2\,\mathrm{d}k \sum_{Jll'l''}[(-\mathrm{i})^l C_{l'}\, j_l(kr)$$

$$\times\, \mathscr{U}^{Js}_{l'l''}(r')v^{Js}_{l''l}(r')]\,(l'0\,sm|Jm)(lm-m'sm'|Jm)Y_l^{m-m'}(\hat{\boldsymbol{r}}, \hat{\boldsymbol{r}}') \qquad (4.18)$$

where C_l is given below (3.1). For singlet states $l' = l'' = J$, and a simple equation results

$$\langle \boldsymbol{r}'|t|\boldsymbol{r}\rangle_{\text{singlet}} = \frac{4\pi}{(2\pi)^3}\int k^2\,\mathrm{d}k \sum_l (2l+1)j_l(kr)\mathscr{U}^{l0}_{ll}(r')v^{l0}_{ll}P_l^0(\hat{\boldsymbol{r}}, \hat{\boldsymbol{r}}'). \qquad (4.19)$$

Equation (18) for triplet states may be cast in a form in which its spin and angular momentum dependence is clearly exhibited. For this purpose methods similar to those used in phenomenological construction of two particle scattering matrix (K15, W8) may be conveniently applied. Introducing the vectors

$$\boldsymbol{e}_- = \frac{\boldsymbol{r}'}{r'} - \frac{\boldsymbol{r}}{r}; \quad \boldsymbol{e}_\perp = \frac{\boldsymbol{r}\times\boldsymbol{r}'}{rr'}; \quad \boldsymbol{e}_n = \boldsymbol{e}_-\times\boldsymbol{e}_\perp \qquad (4.20)$$

one has the general form of the t-matrix for triplet state from invariance considerations

$$\langle \boldsymbol{r}'|t|\boldsymbol{r}\rangle_{\text{triplet}} = A + 2\mathrm{i}B(\boldsymbol{S}\cdot\boldsymbol{e}_\perp) + C(\boldsymbol{\sigma}_1\cdot\boldsymbol{e}_-)(\boldsymbol{\sigma}_2\cdot\boldsymbol{e}_-) + D(\boldsymbol{\sigma}_1\cdot\boldsymbol{e}_n)(\boldsymbol{\sigma}_2\cdot\boldsymbol{e}_n). \qquad (4.21)$$

where $\boldsymbol{S} = \frac{1}{2}(\boldsymbol{\sigma}_1 + \boldsymbol{\sigma}_2)$.

Each term in this equation is a 3 × 3 matrix in the space spanned by the triplet spin functions $\chi_1^m (m = \pm 1, 0)$. A, B, C and D are functions of $\boldsymbol{r}$ and $\boldsymbol{r}'$. For example A is multiplied by the unit matrix and the matrix for the coefficient of B is

$$(\boldsymbol{S}\cdot\boldsymbol{e}_\perp) = \begin{pmatrix} e_\perp^0 & e_\perp^+ & 0 \\ e_\perp^- & 0 & e_\perp^+ \\ 0 & e_\perp^- & -e_\perp^0 \end{pmatrix} \begin{matrix} m \\ 1 \\ 0 \\ -1 \end{matrix} \tag{4.22}$$

where

$$e_\perp^0 = (\boldsymbol{e}_\perp)_z; \quad \sqrt{2}\, e_\perp^\pm = (\boldsymbol{e}_\perp)_x \pm \mathrm{i}(\boldsymbol{e}_\perp)_y = \sin\theta\, \mathrm{e}^{\pm \mathrm{i}\varphi}.$$

Hence if we disregard the C and D terms

$$A(\boldsymbol{r}, \boldsymbol{r}') = \tfrac{1}{3} \sum_m \langle \boldsymbol{r}'\, 1\, m\, |t|\, \boldsymbol{r}\, 1\, m\rangle \tag{4.23}$$

and

$$B(\boldsymbol{r}, \boldsymbol{r}') = \frac{\sqrt{2}}{4\mathrm{i}} [\langle \boldsymbol{r}'\, 1 - 1\, |t|\, \boldsymbol{r}\, 1\, 0\rangle + \langle \boldsymbol{r}'\, 1\, 0\, |t|\, \boldsymbol{r}\, 1\, 1\rangle] \operatorname{cosec}\theta\, \mathrm{e}^{-\mathrm{i}\varphi}. \tag{4.24}$$

Then using the orthogonality relation of Clebsch-Gordan coefficients and (18) for t-matrix ($l = l'$ and change l'' to l')

$$A(\boldsymbol{r},\boldsymbol{r}') = \frac{4\pi}{3(2\pi)^3} \int k^2\, \mathrm{d}k [\sum_{Jll'} (2J+1) j_l(kr) \mathscr{U}_{ll'}^{J1}(r') v_{l'l}^{J1}(r') P_l^0(\hat{\boldsymbol{r}}, \hat{\boldsymbol{r}}')]. \tag{4.25}$$

Similarly

$$B(\boldsymbol{r},\boldsymbol{r}') = \frac{1}{2\sqrt{2}} \frac{4\pi}{(2\pi)^3} \int \Big[k^2\, \mathrm{d}k \sum_{Jll'l''} [\quad] \{(l'\,0\,1\,0\,|\,J\,0)(l\,1\,1-1\,|\,J\,0)$$
$$+ (l'\,0\,1\,1\,|\,J\,1)(l\,1\,1\,0\,|\,J\,1)\} \Big] Y_l^1(\hat{\boldsymbol{r}}, \hat{\boldsymbol{r}}') \operatorname{cosec}\theta\, \mathrm{e}^{-\mathrm{i}\varphi}, \tag{4.26}$$

where [ ] stands for the contents of these brackets in (18). This can be converted to an $\boldsymbol{L}\cdot\boldsymbol{S}$ term by noting that

$$Y_l^1(\mu)\, \mathrm{e}^{-\mathrm{i}\varphi} \operatorname{cosec}\theta = [l(l+1)]^{-\frac{1}{2}} \frac{\mathrm{d}}{\mathrm{d}\mu} Y_l^0(\mu); \quad \mu = \cos\theta \tag{4.27}$$

and

$$\frac{1}{\mathrm{i}} (\boldsymbol{S}\cdot\boldsymbol{e}_+) \frac{\mathrm{d}}{\mathrm{d}\mu} f(r, r', \mu) = \frac{1}{\mathrm{i}} (\boldsymbol{S}\cdot\boldsymbol{r} \times \boldsymbol{\nabla} f(r, r', \mu))$$
$$= \boldsymbol{L}\cdot\boldsymbol{S} f(r, r', \mu) \tag{4.27'}$$

where

$$\boldsymbol{L} \equiv \boldsymbol{L}_{12} = \boldsymbol{r}_{12} \times \boldsymbol{p}_{12} = \tfrac{1}{2}(\boldsymbol{r}_1 \times \boldsymbol{p}_1 + \boldsymbol{r}_2 \times \boldsymbol{p}_2 - \boldsymbol{r}_1 \times \boldsymbol{p}_2 - \boldsymbol{r}_2 \times \boldsymbol{p}_1)$$

so that the coefficient of $\boldsymbol{L}\cdot\boldsymbol{S}$ term, $B'(\boldsymbol{r}, \boldsymbol{r}')$, in the matrix becomes, from (26)

$$B'(\boldsymbol{r}, \boldsymbol{r}') = \frac{1}{\sqrt{2}} \frac{4\pi}{(2\pi)^3} \int \Big[\quad \Big] [l(l+1)]^{-\frac{1}{2}} Y_l^0(\hat{\boldsymbol{r}}, \hat{\boldsymbol{r}}').$$

For practical computation the sum of J, l, l', l'' is limited by considerations mentioned at the end of the previous section and $\Big[\quad \Big]$ stands for the terms contained in such brackets in (26).

4a (i). Hard core contributions

The t-matrix elements may be written as

$$\langle \boldsymbol{r}' | t | \boldsymbol{r} \rangle = \langle \boldsymbol{r}' | t | \boldsymbol{r} \rangle_{\text{core}} + \langle \boldsymbol{r}' | t | \boldsymbol{r} \rangle_a, \tag{4.28}$$

where the separation is achieved by writing [42]

$$v = v_a + v_{\text{core}} \tag{4.28$'$}$$

and making the replacement

$$v_{\text{core}}(r) \mathscr{U}_{ll'}^{Js}(r) = \lambda_{ll'}^{Js} \delta(r - r_c) \tag{4.29}$$

where r_c is the core radius.

4a (ii). The t-matrix for S-state interaction: an approximate form (B17)

For singlet S-states under the approximation of replacing $\mathscr{U}(r')$ by $j_0(kr')$ (first Born approximation) in the expression (3.28) for λ and neglecting the slow variation of Green's function on k, the core contribution is

$$\langle \boldsymbol{r}' | t | \boldsymbol{r} \rangle_{\text{core}} = -\frac{\delta(r' - r_c)\delta(r - r_c)}{(4\pi r_c^2)^2 G_0(r_c, r_c)} - \frac{G(r_c, r')}{G(r_c, r_c)} \frac{v_a(r')}{4\pi r_c^2} \delta(r - r_c). \tag{4.30}$$

For the same states, using the second Born approximation for $\mathscr{U}$, the

[42] Moszkowski and Scott (M7, M8) have considered a different splitting of the potential which gives rise to certain other advantages in computational procedure.

contribution of the attractive part comes out to be

$$\langle \boldsymbol{r}' |t| \boldsymbol{r} \rangle_a = \frac{\delta(r - r')}{4\pi r^2} v_a - \delta(r' - r_c) \frac{v_a(r)}{4\pi r_c^2} \frac{G_0(r, r_c)}{G_0(r_c, r_c)} . \quad (4.31)$$

Here the t-matrix is seen to possess "lines of strong repulsion" along $r' = r_c$ and $r = r_c$, as well as repulsion "spikes" at $r = r' = r_c$. Apart from these the t-matrix is local and behaves like $v_a(r)/r^2$.

The effect of more accurate calculations (B17) is to "smooth out" the δ-function singularities. There then arise off-diagonal regions of strong repulsion. It turns out (B17), however, that beyond a distance of about one fermi the t-matrix is effectively diagonal. The quantity

$$v_{\text{eff}}(r) = \int \langle \boldsymbol{r}' |t| \boldsymbol{r} \rangle \, \mathrm{d}\boldsymbol{r}'$$

is for large r equal to $v_a(r)$.

A measure of the correlation distance may be obtained by noting the distance at which $v_{\text{eff}}(r)$ becomes indistinguishable from $v_a(r)$. This quantity is comparable to the "healing" distance, which was defined in terms of the two particle wave function.

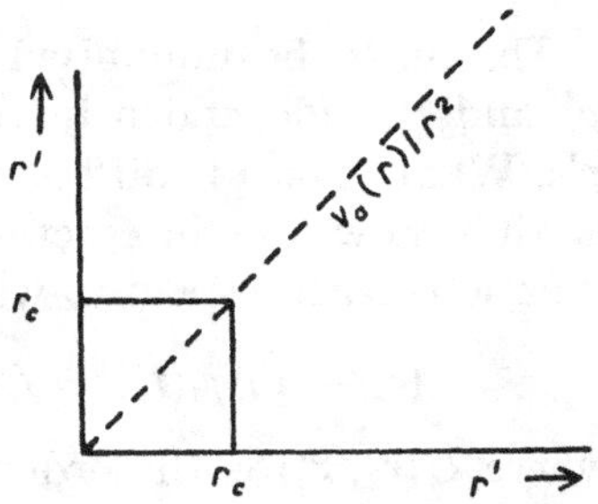

Fig. 31. Configuration space representation of the non-local t-matrix, Eqs. (IV. 4.30) and (31). There are sharp infinitely high repulsive edges at the lines $r_c = r$ and $r_c = r'$. A more careful evaluation of the t-matrix gives rise to off-diagonal regions of repulsion which are strongest near the edges (B17).

4b. *Single particle potential*

This quantity is of special interest for nuclei. For a non-local t-matrix the potential will also be non-local. In this section a derivation of the potential in terms of a Hartree-Fock variational principle (e.g., S11) [43] is given to pave the way for discussion of single

[43] Overhauser (O1) has pointed out that there may exist solutions of the many-fermion problem which correspond to giant matter-density or spin-density fluctuations and have energies less than the usual uniform Hartree-Fock solutions. The existence of such solutions in case of *sufficiently strong* interaction was recognised much earlier by Wigner (W9). Gross (G6) has pointed

particle energies to be given in the next section. The effective interaction is expressed through the t-matrix. If for simplicity we neglect the momentum non-conserving terms the total energy is given by

$$\mathscr{E} = E = \sum_i \int \varphi_i^*(\boldsymbol{r}_1) \frac{p_i^2}{2M} \varphi_i(\boldsymbol{r}_1)\, \mathrm{d}\boldsymbol{r}_1$$

$$+ \tfrac{1}{2} \sum_i \sum_j \int [\varphi_i^*(\boldsymbol{r}_1)\varphi_j^*(\boldsymbol{r}_2) \langle \boldsymbol{r}_1\boldsymbol{r}_2|t|\boldsymbol{r}_1'\boldsymbol{r}_2'\rangle$$

$$\times \{\varphi_i(\boldsymbol{r}_1')\varphi_j(\boldsymbol{r}_2') - \text{exch.}\}]\, \mathrm{d}\boldsymbol{r}_1\, \mathrm{d}\boldsymbol{r}_2\, \mathrm{d}\boldsymbol{r}_1'\, \mathrm{d}\boldsymbol{r}_2'. \quad (4.32)$$

This is to be minimised w.r.t. the single particle wave functions φ_i^* and φ_i independently to obtain the differential equation for the φ's. What makes it different from the usual Hartree-Fock calculation is that now the interaction matrix t itself depends on the wave functions with respect to which variation is made. One gets [44]

$$(\epsilon_i - T_i)\varphi_i(\boldsymbol{r}_1) = \int U(\boldsymbol{r}_1, \boldsymbol{r}_1')\varphi_i(\boldsymbol{r}_1')\, \mathrm{d}\boldsymbol{r}_1' + U_R(\boldsymbol{r}_1)\varphi_i(\boldsymbol{r}_1), \quad (4.33)$$

where $U(\boldsymbol{r}_1, \boldsymbol{r}_1')$ is the ordinary non-local potential to be discussed in detail in this section and $U_R(\boldsymbol{r}_1)$, the rearrangement potential, arises from the variation of t-matrix w.r.t. to the wave function,

$$U_R(\boldsymbol{r}_1)\varphi_i(\boldsymbol{r}_1) = \tfrac{1}{2} \sum_{jk} \int \varphi_j^*(\boldsymbol{r}_2)\varphi_k^*(\boldsymbol{r}_3) \left[\frac{\delta\langle \boldsymbol{r}_2\boldsymbol{r}_3|t|\boldsymbol{r}_2'\boldsymbol{r}_3'\rangle}{\delta\varphi_i^*(\boldsymbol{r}_1)} \right]$$

$$\times \{\varphi_j(\boldsymbol{r}_2')\varphi_k(\boldsymbol{r}_3') - \varphi_j(\boldsymbol{r}_3')\varphi_k(\boldsymbol{r}_2')\}\, \mathrm{d}\boldsymbol{r}_2\, \mathrm{d}\boldsymbol{r}_3\, \mathrm{d}\boldsymbol{r}_2'\, \mathrm{d}\boldsymbol{r}_3'. \quad (4.34)$$

out that the existence of such states is a fairly common phenomenon in the general many-body problem. Overhauser's conjecture, if true, would mean that the discussions given here will not generally apply to the ground state of nuclear matter. However at present there is some doubt whether Overhauser's conjecture based on a one-dimensional example can apply to the three-dimensional fermion systems that occur in nature (K7, B24). A related problem is to find whether there are any unstable solutions or whether the Hartree-Fock solutions themselves are stable. General criteria for the stability of the latter have been given by Pomeranchuk (P3) and Thouless (T7). Unstable modes in nuclear matter were discussed by Glassgold, Heckrotte and Watson (G7) and their connection to the Hartree-Fock method, and results of Overhauser and of Kohn and Nettle (K7) were discussed by Brout (B24), see also (H10).

[44] Since this has been derived from (4.32) it can only be approximate (Ch. III, sec. 5). The single particle potential may also be defined in terms of diagrams where higher order effects can be explicitly included (H4, S12).

The U_R terms affect the total energy only indirectly through the changes in model wave functions and hence the t-matrix. In the nuclear matter case the wave functions are not altered so that U_R has no effect on the total energy.

In the next section we discuss the definitions of single particle energies and show that for the nuclear matter problem U_R may be approximated by a constant $\simeq 20\%$ of the main potential term. The explicit form of U_R will be discussed in connection with nuclei (§ 6).

We devote the rest of this section to a reduction of the main potential in terms of the t-matrix. The ordinary non-local potential $U(\boldsymbol{r}, \boldsymbol{r}')$ is obtained by variation of the potential energy in (32) w.r.t. φ_i^* while neglecting the dependence of the t-matrix itself on φ_i^* ((33) and (34)). $U(\boldsymbol{r}, \boldsymbol{r}')$ is given by

$$\int U(\boldsymbol{r}_1, \boldsymbol{r}_1')\varphi_i(\boldsymbol{r}_1')\, d\boldsymbol{r}_1' = \sum_j \int \varphi_j^*(\boldsymbol{r}_2)[\langle \boldsymbol{r}_1\boldsymbol{r}_2|t|\boldsymbol{r}_1'\boldsymbol{r}_2'\rangle \varphi_j(\boldsymbol{r}_2')\varphi_i(\boldsymbol{r}_1') - \langle \boldsymbol{r}_1\boldsymbol{r}_2|t|\boldsymbol{r}_1'\boldsymbol{r}_2'\rangle \varphi_j(\boldsymbol{r}_1')\varphi_i(\boldsymbol{r}_2')]\, d\boldsymbol{r}_1'\, d\boldsymbol{r}_2'\, d\boldsymbol{r}_2. \tag{4.35}$$

At this point it should be noted that if the two particles ($\boldsymbol{r}_1$ and $\boldsymbol{r}_2$) are of different types there is no requirement of anti-symmetry and hence for this case exchange terms do not appear. When the particles are of the same type the interaction can take place only in the singlet-even or triplet-odd states. The t-matrix can then be split into corresponding parts by use of projection operators

$$\langle \boldsymbol{r}_1\boldsymbol{r}_2|t|\boldsymbol{r}_1'\boldsymbol{r}_2'\rangle = \langle \boldsymbol{r}_1\boldsymbol{r}_2|t|\boldsymbol{r}_1'\boldsymbol{r}_2'\rangle_{s,e}\Lambda_s + \langle \boldsymbol{r}_1\boldsymbol{r}_2|t|\boldsymbol{r}_1'\boldsymbol{r}_2'\rangle_{t,o}\Lambda_t, \tag{4.36}$$

where the spin dependence resides entirely in the projection operators:

$$\Lambda_s = \tfrac{1}{4}(1 - \boldsymbol{\sigma}_1\cdot\boldsymbol{\sigma}_2); \qquad \Lambda_t = \tfrac{1}{4}(3 + \boldsymbol{\sigma}_1\cdot\boldsymbol{\sigma}_2), \tag{4.37}$$

the quantities $(\ )_{s,e}$ and $(\ )_{t,o}$ being not spin dependent.

The exchange part in Eq. (35) may be written

$$-\langle \boldsymbol{r}_1\boldsymbol{r}_2|t|\boldsymbol{r}_2'\boldsymbol{r}_1'\rangle \varphi_j(\boldsymbol{r}_2')\varphi_i(\boldsymbol{r}_1')$$

where exchange in the t-matrix involves an exchange of spin-variables also (all coordinates); with the projection operators the exchange term becomes

$$-[\langle \boldsymbol{r}_1\boldsymbol{r}_2|t|\boldsymbol{r}_2'\boldsymbol{r}_1'\rangle_{s,e}\Lambda_s + \langle \boldsymbol{r}_1\boldsymbol{r}_2|t|\boldsymbol{r}_2'\boldsymbol{r}_1'\rangle_{t,o}\Lambda_t]P_\sigma\varphi_j(\boldsymbol{r}_2')\varphi_i(\boldsymbol{r}_1'), \tag{4.38}$$

where P_σ is the spin exchange operator

$$P_\sigma = \tfrac{1}{2}(1 + \boldsymbol{\sigma}_1\cdot\boldsymbol{\sigma}_2). \tag{4.39}$$

We have the relations

$$\Lambda_s P_\sigma = -\Lambda_s; \qquad \Lambda_t P_\sigma = +\Lambda_t. \tag{4.39'}$$

The exchange of space coordinates of the two particles is equivalent to parity operation since the t-matrix is basically a function of relative coordinates $\boldsymbol{r}_{12}$ and $\boldsymbol{r}'_{12}$. The $(\;)_{s,e}$ and $(\;)_{t,o}$ terms have even and odd parity respectively, hence the exchange is equivalent to a change of sign giving a term equal to the direct one for *like* particles. The contributions for neutrons and protons may be separated, remembering that there are no exchange terms for unlike particles.

The single particle potential for neutrons is thus given by

$$U_n(\boldsymbol{r}_1, \boldsymbol{r}'_1) = \tfrac{1}{2} \sum_{j(\text{neutrons})} \int \varphi_j^*(\boldsymbol{r}_2)[t_{s,e} + 3t_{t,0}]\varphi_j^*(\boldsymbol{r}'_2)\, \mathrm{d}\boldsymbol{r}_2\, \mathrm{d}\boldsymbol{r}'_2$$

$$+ \tfrac{1}{4} \sum_{k(\text{protons})} \int \varphi_k^*(\boldsymbol{r}_2)[t_{s,e} + 3t_{t,e} + t_{s,o} + 3t_{t,0}]\varphi_k(\boldsymbol{r}'_2)\, \mathrm{d}\boldsymbol{r}_2\, \mathrm{d}\boldsymbol{r}'_2; \qquad (4.40)$$

$$t \equiv \langle \boldsymbol{r}_1\boldsymbol{r}_2 | t | \boldsymbol{r}'_1\boldsymbol{r}'_2 \rangle.$$

Or, more simply

$$U_n(\boldsymbol{r}_1, \boldsymbol{r}'_1) = \sum_j C_{ij} \int \varphi_j^*(\boldsymbol{r}_2) \langle \boldsymbol{r}_1\boldsymbol{r}_2 | t | \boldsymbol{r}'_1\boldsymbol{r}'_2 \rangle \varphi_j(\boldsymbol{r}'_2)\, \mathrm{d}\boldsymbol{r}_2\, \mathrm{d}\boldsymbol{r}'_2, \qquad (4.40')$$

where C_{ij} is the appropriate weight factor which, if necessary, may be written in terms of the isobaric and mechanical spin variables of i and j. A similar expression may also be written down for the proton potential.

4b (i). Spin-orbit part of single particle potential (B17, N3, S10)

Before the advent of t-matrix theory it was realised by several workers that the spin-orbit part of the single particle potential which is so important for the shell model can arise in first order from the spin-orbit part of the two-body potential (B39, H9) and/or in second order from the tensor part (F15, K19) – the radial parts in both cases being assumed well behaved. The basic physical ideas remain the same but in the present case the spin-orbit part is seen to arise from the B-part of the t-matrix and can contain in it terms arising from both tensor and spin orbit parts of the two particle interaction

$$v = v_c + S_{12}v_T + v_{LS}\boldsymbol{L}\cdot\boldsymbol{S}. \qquad (4.41)$$

The contributions arise directly through the occurrence of the v_{LS} and v_T terms in the t-matrix and indirectly through their influence on $\mathscr{U}_{ll'}^{Js}$ wave functions.

To obtain the single particle $\boldsymbol{L}_1 \cdot \boldsymbol{S}_1$ term of the potentials one has to reduce the two-body $\boldsymbol{L}_{12} \cdot \boldsymbol{S}_{12}$ term. First note that

$$\boldsymbol{p}_2 B'(\boldsymbol{r}_{12}, \boldsymbol{r}'_{12}) = -\boldsymbol{p}_1 B'(\boldsymbol{r}_{12}, \boldsymbol{r}'_{12})$$

and hence

$$\boldsymbol{L}_{12} B' = (\boldsymbol{r}_{12} \times \boldsymbol{p}_{12}) B' = (\boldsymbol{r}_{12} \times \boldsymbol{p}_1) B'. \tag{4.42}$$

The single particle potential is

$$\begin{aligned}\sum_j C_{ij} \int \varphi_j^*(\boldsymbol{r}_2)(\boldsymbol{r}_{12} \times \boldsymbol{p}_1) B'(\boldsymbol{r}_{12}, \boldsymbol{r}'_{12}) \delta(\boldsymbol{R} - \boldsymbol{R}') \varphi_j(\boldsymbol{r}'_2)\, \mathrm{d}\boldsymbol{r}_2\, \mathrm{d}\boldsymbol{r}'_2 \\ = \sum_j C_{ij} \int \varphi_j^*(\boldsymbol{r}_2)(\boldsymbol{r}_{12} \times \boldsymbol{p}_1) B'(\boldsymbol{r}_{12}, \boldsymbol{r}_{12} - 2(\boldsymbol{r}_1 - \boldsymbol{r}'_1)) \\ \times \varphi_j(\boldsymbol{r}_2 + \boldsymbol{r}_1 - \boldsymbol{r}'_1)\, \mathrm{d}\boldsymbol{r}_2.\end{aligned} \tag{4.43}$$

Note that $\boldsymbol{p}_1$ also acts on φ_j because the presence of the δ-function introduces an $\boldsymbol{r}_1$ dependence on φ_j. If B' were local, then the only non-zero contribution would be from regions where $\boldsymbol{r}_1 = \boldsymbol{r}'_1$. Hence if the single particle wave functions do not change appreciably over the range of non-locality of the *t*-matrix, more accurately of its $\boldsymbol{L} \cdot \boldsymbol{S}$ part, then $\boldsymbol{p}_1$ may be taken out of the integral sign. For the same reason one may write inside the integral

$$\boldsymbol{r}_{12} = \boldsymbol{r}_1 - \boldsymbol{r}_2 \cong \boldsymbol{r}_1 \left(1 - \frac{\boldsymbol{r}_1 \cdot \boldsymbol{r}_2}{r_1^2}\right).$$

Finally, the two parts of the single particle potential, the central and the spin-orbit one, become

$$\langle \boldsymbol{r}_1 | U_i^{(c)} | \boldsymbol{r}'_1 \rangle = \sum_j C_{ij} \int \mathrm{d}\boldsymbol{r}_2\, \mathrm{d}\boldsymbol{r}'_2\, \varphi_j^*(\boldsymbol{r}_2) \langle \boldsymbol{r}_{12} | t | \boldsymbol{r}'_{12} \rangle_{\text{central}}\, \varphi_j(\boldsymbol{r}'_2) \tag{4.44}$$

$$\begin{aligned}&\langle \boldsymbol{r}_1 | U_i^{LS} | \boldsymbol{r}'_1 \rangle \\ &= \sum_j C_{ij} \int \mathrm{d}\boldsymbol{r}_2 \mathrm{d}\boldsymbol{r}'_2 \varphi_j^*(\boldsymbol{r}_2) \langle \boldsymbol{r}_{12} | t | \boldsymbol{r}'_{12} \rangle_{LS} \left(1 - \frac{\boldsymbol{r}_1 \cdot \boldsymbol{r}_2}{r_1^2}\right) \cdot \varphi_j(\boldsymbol{r}'_2).\end{aligned} \tag{4.45}$$

It was shown in ref. (B17) and (S10) that the last expression gives rise to a Thomas type term in the first approximation, but this approximation is no longer considered good (B35). (See ref. (C7) for an alternative treatment.) A connected problem is the spin orbit interaction in the optical model which causes polarisation of the incident particles (K15, and references quoted therein).

5. SINGLE PARTICLE ENERGIES

The single particle energies, abbreviated as SPE in the following and defined in (1.3) have so far been used only to provide the energy denominators for the t-matrix equation (1.1) which, however, remains unaltered even if we add an arbitrary constant to the single particle energies. This invariance of the t-matrix and hence also of the total energy under the translations of the origin of SPE scale indicates that the present values of SPE may have no direct connection with the experimentally observed quantities like separation energies of particles and the average binding energies. It is desirable that the SPE should be defined in such a way that a connection exists. In Ch. II, sec. 4g (theorem of Hugenholtz and Van Hove (H4)), the SPE is defined in terms of diagrams and then such a connection can be shown to hold rigorously (see also (B26, K1, K16, M5, V1)).

In terms of t-matrix vertices the total energy is represented by the sequence of diagrams in Fig. 32a. It is important to note that no second order diagrams occur. It appears that the third order terms are small and one may presume that the convergence is good. On the

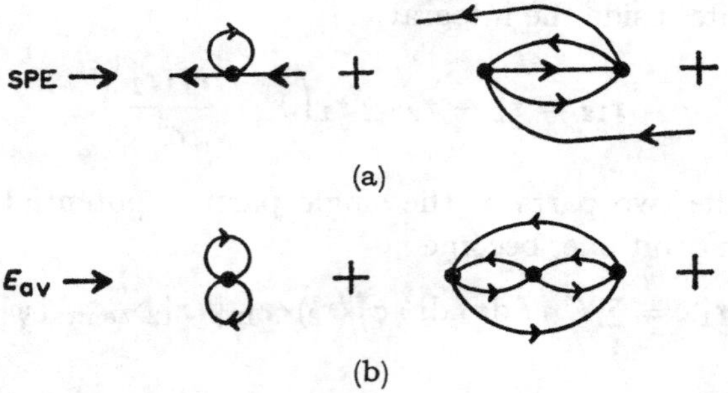

Fig. 32. Series of t-vertex diagram for (a) total energy and (b) single particle energy. Note that the second order diagram does not occur in (a).

other hand the SPE is given by means of the diagrams given in Fig. 32b. The occurrence of second order t-terms in expressions for SPE is significant.

In order that the first order terms be an adequate approximation of E_{Av}, the magnitude of second order t-matrix terms is irrelevant, only third order and beyond are required to be small. Equation (1.3) for

SPE corresponds to taking the first order terms in the true expansion of SPE. For it to be a good approximation the second order terms must be small. If this is not the case, then the theorem of Hugenholtz and Van Hove (H4) will not be satisfied by these approximate SPE and average energies. In the work of Brueckner and Gammel (B19) this actually happens and a discrepancy of about 12 MeV is observed: $\epsilon_F \simeq 27.5$ MeV, $E_{Av} \simeq 15.5$ MeV. It is gratifying that the magnitude [45] of neglected second order terms for ϵ_F is also $\simeq$ 12 MeV (H4).

The consequences of this discrepancy are presumably not quite so disastrous as may first seem. The reasons for this are twofold. Firstly, the t-matrix is invariant under translation of the energy scale for single particle energies, so that if the dependence of the higher terms of single particle energies on momentum is weak, their effect on the t-matrix would be small even if their magnitude is quite large. Secondly, the t-matrix does not seem to be sensitive to further refinements on the propagator as the experience (B19) with the off-the-energy-shell propagation shows. A complete discussion of this point, the use of SPE in propagators, is connected also with the very question of higher order corrections. It is a complex situation and is difficult to analyse.

Brueckner (B26) has proposed a simple redefinition of the SPE which approximately satisfies the theorem of Hugenholtz and Van Hove. He writes

$$\epsilon'(k_i) = \frac{k_i^2}{2M} + \tfrac{1}{4} \int \mathrm{d}\boldsymbol{k}_j\, f(ij) + \tfrac{1}{3}\Delta = \frac{k_i^2}{2M} + W(k_i) \tag{5.1}$$

where

$$\tfrac{1}{4} \int \mathrm{d}\boldsymbol{k}_j\, f(ij) = U(k_i) = \tfrac{1}{4} \frac{\Omega}{(2\pi)^3} \int \sum_{\substack{\text{spins}\\(ij)}} \langle ij|t|ij - ji\rangle\, \mathrm{d}\boldsymbol{k}_j \tag{5.2}$$

and the constant Δ is determined by requiring that the new Fermi-energy is equal to E_{Av}

$$\epsilon'(k_F) = E_{Av}. \tag{5.3}$$

[45] In fact there is no general agreement on the numerical magnitude of almost any quantity arising from this theory. The estimates of rearrangement energy have varied from 6 to 20 MeV (see footnote 11 of Moszkowski and Scott (M7) for a list of references). A more recent work of Brueckner *et al.* on this topic is (B34).

Therefore,

$$\begin{aligned}\tfrac{1}{3}\Delta &= E_{\mathrm{Av}} - \epsilon(k_{\mathrm{F}})\\ &= \tfrac{3}{5}\frac{k_{\mathrm{F}}^2}{2M} + U_{\mathrm{Av}} - \frac{k_{\mathrm{F}}^2}{2M} - U(k_{\mathrm{F}})\\ &= U_{\mathrm{Av}} - U(k_{\mathrm{F}}) - \tfrac{2}{5}\frac{k_{\mathrm{F}}^2}{2M},\end{aligned} \tag{5.4}$$

where

$$U_{\mathrm{Av}} = \tfrac{1}{2}\frac{\Omega_0}{(2\pi)^3}\int \mathrm{d}\boldsymbol{k}_i\,\mathrm{d}\boldsymbol{k}_j\, f(ij),$$

Ω_0 is the volume per particle.

The relation can be further simplified by using the condition of saturation of density ($\sim k_{\mathrm{F}}^3$), in the form that E_{Av} is independent of k_{F}, i.e.

$$k_{\mathrm{F}}\left(\frac{\partial E_{\mathrm{Av}}}{\partial k_{\mathrm{F}}}\right) = 0. \tag{5.5}$$

Using the relation

$$k_{\mathrm{F}}(\partial\Omega_0/\partial k_{\mathrm{F}}) = -3\Omega_0,\ (\because\ \Omega_0 \sim k_{\mathrm{F}}^{-3}) \tag{5.6}$$

and

$$\frac{4\Omega_0}{(2\pi)^3}\,\tfrac{4}{3}\pi k_{\mathrm{F}}^3 = 1 \tag{5.6'}$$

(the factor 4 from spin and charge states) in the differentiated form of E_{Av}, we have

$$0 = \tfrac{6}{5}\frac{k_{\mathrm{F}}^2}{2M} - 3U_{\mathrm{Av}} + 2\int 4\pi k_{\mathrm{F}}^3\,\tfrac{1}{2}\frac{\Omega_0}{(2\pi)^3}\,\mathrm{d}\boldsymbol{k}_j\, f(ij)\Big|_{k_i=k_{\mathrm{F}}} + \tfrac{1}{2}\frac{\Omega_0}{(2\pi)^3}\int \mathrm{d}\boldsymbol{k}_i\,\mathrm{d}\boldsymbol{k}_j\, k_{\mathrm{F}}\frac{\partial f(ij)}{\partial k_{\mathrm{F}}}.$$

Or

$$\tfrac{2}{5}\frac{k_{\mathrm{F}}^2}{2M} - U_{\mathrm{Av}} + U(k_{\mathrm{F}}) + \tfrac{1}{3}\tfrac{1}{2}\frac{\Omega_0}{(2\pi)^3}\int \mathrm{d}\boldsymbol{k}_i\,\mathrm{d}\boldsymbol{k}_j\, k_{\mathrm{F}}\frac{\partial f(ij)}{\partial k_{\mathrm{F}}} = 0.$$

Hence finally,

$$\Delta = \tfrac{1}{2}\frac{\Omega_0}{(2\pi)^3}\int \mathrm{d}\boldsymbol{k}_i\,\mathrm{d}\boldsymbol{k}_j\, k_{\mathrm{F}}\frac{\partial f(ij)}{\partial k_{\mathrm{F}}}. \tag{5.7}$$

This may be looked upon as the momentum average of the higher terms in the t-matrix expansion of SPE.

In terms of the new potential $W(k_i)$, we have

$$U_{\mathrm{Av}} = \frac{1}{2N} \sum_i \left(W(k_i) - \tfrac{1}{3}\Delta\right) \tag{5.8}$$

where N is the number of particles, equation (3.19) still holds for E_{Av}.

The Δ term is quite important, giving a correction of $\simeq 20\%$ to $U(k_{\mathrm{F}})$. It is clear that these corrections will require more careful attention in problems such as that of the optical model where the momentum averaging of higher terms may not be such a good approximation.

Finally, it should be noted that Δ is connected to the rearrangement potential U_R. In the approximation of a constant $U_R(\boldsymbol{r}) = \overline{U}_R$

$$\overline{U}_R \simeq \tfrac{1}{3}\Delta. \tag{5.9}$$

More sophisticated treatments of rearrangement energy have been given by Brueckner *et al.* (B34, B36) and others (K16, M5, V1).

6. APPLICATION TO NUCLEI [46]

So far the general theory has been considered and its application to the infinite medium (nuclear matter) has been used as the stock example for illustrations. This idealisation permits the use of plane wave functions and a continuous eigenvalue spectrum for the model system. This is the source of some major simplifications: (1) Wave function self-consistency is always obtained, since the perturbed wave functions also have to be plane waves for uniform infinite systems. (2) The rearrangement potential U_R has no influence on the wave functions for the same reason. (3) A large class of matrix-elements vanishes in view of momentum conservation

$$(\boldsymbol{l}_1\boldsymbol{l}_2|v|\boldsymbol{l}_3\boldsymbol{l}_4) = (\boldsymbol{l}_1\boldsymbol{l}_2|t|\boldsymbol{l}_3\boldsymbol{l}_4) = 0$$

if

$$(\boldsymbol{l}_1 - \boldsymbol{l}_2) - (\boldsymbol{l}_3 - \boldsymbol{l}_4) \neq 0$$

and

$$(\boldsymbol{l}_1|p^2|\boldsymbol{l}_2) = 0 \text{ if } \boldsymbol{l}_1 \neq \boldsymbol{l}_2.$$

This leads to a considerable reduction in computational work.

[46] The term 'finite nuclei' has often been used in the literature to refer to the nuclei themselves as distinct from nuclear matter – a hypothetical construct. The adjective is actually superfluous and in interest of clarity we shall not use it.

(4) Actually the equations of the theory are 'summation equations' but the use of a continuous spectrum makes them into integral equations on which, in certain approximations, the mathematical methods for solving the latter may be used.

In applying the theory to nuclei the main source of complication is the requirement of wave function self-consistency (breakdown of (1) above). For the same reason the rearrangement potential, now having a finite extension in space, has an important effect on the wave functions. Self-consistency is needed only to ensure that the perturbation theory (for the final set of wave functions and t-matrices) is rapidly convergent. It follows then that self-consistency is not so absolutely necessary if we can be assured of the convergence in some other way. This observation is useful since it can save us much computational labour in appropriate circumstances. We need to satisfy the self-consistency condition only approximately.

Within the framework of Brueckner theory there have been three major calculations for nuclei. First is the method outlined by Brueckner, Gammel and Weitzner (abbreviated as BGW, ref. B17), with corrections later noted by Brueckner and Goldman (B25). This method has been applied to O^{16}, Ca^{40} and Zr^{90} (B18, B35). The second is by Eden and co-workers (E4) concerning O^{16} and involves certain important differences. The third is by Mang and Wild (M2) concerning the three and four body problem where Brenig's formulation of the many body problem (B8, Ch. V, sec. 3) is used. All these works report excellent agreement with binding energies etc., and in all these works there are several unevaluated approximations. It is not possible to comment on the effect of these approximations without detailed numerical work. The results of different works cannot be compared because they use different variants of the theory. Often the conclusions of the same group keep changing with each new calculation. Therefore we shall say very little about the numerical results or the validity of these methods which so often depends on numerical smallness of certain terms. We present in outline the first two of the methods mentioned above. It is to be hoped that in time an agreement will be reached on a definite scheme of calculation so that results of various works may be compared with facility and confidence.

An especial difficulty in treating nuclei arises from the centre of mass motion. The model single particle wave functions have to be taken as wave functions inside a potential of finite extension, fixed in space. The many body model wave functions are determinants formed from these single particle wave functions centred at the origin. In any given state of the (model) system the centre of mass executes a certain motion, the energy of which is included in the model energy. In general this cannot be separated easily and hence there is a spurious contribution to the energy. The same difficulty arises in the shell model theory where it has been shown that for simple harmonic wave functions the centre of mass motion is separable and the spurious states can be removed (E6). Based on the same type of thinking is a suggestion by Bolsterlei and Feenberg (B10) which was later extended to the *t*-matrix theory by Lipkin (L2) for removing the centre of mass motion from the theory of nuclei. This suggestion has been put to use in the work of Mang and Wild (M2) and is briefly described in a following section.

The problem of the single particle potential in nuclei has been considered for instance in ref. (J3) for the optical model and in ref. (C7) for spin-orbit splitting of levels.

6a. *Brueckner-Gammel-Weitzner method (B17)*

The problem of having to solve self-consistently for all the three quantities, i.e., the *t*-matrix, the wave functions and the energies according to the plan sketched in sec. 3e, for nuclei, is formidable. It is perhaps beyond the capacity of currently available fast computers (B17).

The simplifying principle used by BGW is to give up the requirement of complete self-consistency. The *t*-matrix for nuclear matter problems has been calculated earlier by Brueckner and Gammel (B19) and is given as a function of density or of the volume $\frac{4}{3}\pi r_0^3$ occupied by each particle. This *t*-matrix is self-consistent at each density value for an infinite medium. In actual nuclei the density is not uniform but drops off towards the edge. The *t*-matrix of the nucleus thus will also depend on the distance from the centre of the nucleus, a dependence not present in the *t*-matrix for an infinite medium. It is now argued that if the variation in density of the nucleus is small over the distances of the order of the "healing" or "correlation" distance, then the *t*-matrix appropriate for a particular point in the nucleus will be effectively the same as the t-matrix for nuclear matter at a density equal to the density at that point. (This argument denies the existence of long range effects on the t-matrix.) BGW believe that the approximation will still be good even if the density varies linearly over distances of the order of the "healing" distance. The physical reason behind this approximation is as follows: One observes that the exclusion principle requires the energy denominators in the t-matrix equation to be large; on the average nucleons are excited to relative momenta $p \simeq 1.5 p_F$ which at normal nuclear densities correspond to a wave length $\lambdabar \simeq 0.5$ fm of relative motion. If the variation of density is small over distances [47] of this size, then it would be a good enough approximation to replace the actual energy denominators by those appropriate to a Fermi gas (*i.e.*, nuclear matter) at the local density.

With this much justification one takes the coordinate space representation of the t-matrix operator as obtained from nuclear matter studies (B19) and assumes that its expectation value with respect to the model wave functions of a nucleus will be close enough to that of the actual t-operator for that nucleus. In taking the expectation value, the density parameter is also varied so as to conform to the density in the nucleus. The density at the point $\boldsymbol{r}$ is given by

$$\rho(\boldsymbol{r}) = \sum_i |\varphi_i(\boldsymbol{r})|^2. \tag{6.1}$$

The t-matrix is diagonal in the centre of mass coordinates

$$(\boldsymbol{r}_1 \boldsymbol{r}_2 | t(\rho) | \boldsymbol{r}_1' \boldsymbol{r}_2') = \delta(\boldsymbol{R} - \boldsymbol{R}')(\boldsymbol{r}_{12} | t(\rho) | \boldsymbol{r}_{12}') \tag{6.2}$$

and the value of ρ at the position of centre of mass is taken.

The rearrangement potential (B26, B25) is expected to play an important role in the theory of nuclei and may now be written more

[47] The "smoothness" of density distribution given experimentally by Stanford experiments does not necessarily ensure this. One should inquire whether their experiments were sensitive to this type of inhomogeneity and whether they were concerned with averages of the same kind. A study by Drummond (D8) shows that correlations of less than a fermi do not affect the electron scattering at currently available high energies.

explicitly

$$U_R(\boldsymbol{r}_1) = \frac{1}{2\varphi(\boldsymbol{r}_1)} \sum_{jk} \int \varphi_j^*(\boldsymbol{r}_2)\varphi_k^*(\boldsymbol{r}_3) \left[\frac{\delta(\boldsymbol{r}_2\boldsymbol{r}_3|t|\boldsymbol{r}_2'\boldsymbol{r}_3')}{\delta\varphi^*(\boldsymbol{r}_1)}\right] \times \{\varphi_j(\boldsymbol{r}_2')\varphi_k(\boldsymbol{r}_3') - \varphi_j(\boldsymbol{r}_3')\varphi_k(\boldsymbol{r}_2')\}\, d\boldsymbol{r}_2\, d\boldsymbol{r}_3\, d\boldsymbol{r}_2'\, d\boldsymbol{r}_3' \tag{6.3}$$

$$\frac{\delta t}{\delta\varphi^*(\boldsymbol{r}_1)} = \frac{\delta t}{\delta\rho(\boldsymbol{R})} \cdot \frac{\delta\rho(\boldsymbol{R})}{\delta\varphi^*(\boldsymbol{r}_1)} = \frac{\delta t}{\delta\rho}\varphi(\boldsymbol{r}_1)\delta(\boldsymbol{R} - \boldsymbol{r}_1). \tag{6.4}$$

Hence, finally

$$U_R(\boldsymbol{r}_1) = \tfrac{1}{2} \sum_{jk} \int \varphi_j^*(\boldsymbol{r}_2)\varphi_k^*(\boldsymbol{r}_3) \left(\boldsymbol{r}_2\boldsymbol{r}_3 \left| \frac{\delta t}{\delta\rho} \right| \boldsymbol{r}_2'\boldsymbol{r}_3'\right) \delta(\boldsymbol{R} - \boldsymbol{r}_1) \times \{\varphi_j(\boldsymbol{r}_2')\varphi_k(\boldsymbol{r}_3') - \varphi_j(\boldsymbol{r}_3')\varphi_k(\boldsymbol{r}_2')\}\, d\boldsymbol{r}_2\, d\boldsymbol{r}_3\, d\boldsymbol{r}_2'\, d\boldsymbol{r}_3'. \tag{6.5}$$

By leaving out the question of exact self-consistency the method reduces to a Hartree-Fock type of calculation. The single particle wave functions are obtained by solving the Eq. (4.33) with U_R properly taken into account. The binding energy is then calculated using Eq. (1.3) where the expectation value w.r.t. single particle wave function is taken. The coordinate space representation of the t-matrix for infinite nuclear matter is used.

It has been pointed out (K17) that there exists an approximation which is related to the BGW method somewhat as the Fermi-Thomas method is related to the Hartree-Fock. There are corresponding advantages in simplicity of calculation. As in the Fermi-Thomas method one can obtain information on binding energies, density distributions, and distribution of particles in l-levels. The method is specially suited for studying surface phenomena and in that connection has some similarities with earlier semi-empirical methods (S13, W13).

For applying Hartree-Fock methods, equation (4.33) should be reduced in angular and radial parts and BGW further show that the Hartree-Fock problem can be approximately reduced to a differential equation. The details are set out below.

With a slight redefinition U_R can be absorbed in the central part of the potential

$$U'^{(C)}(\boldsymbol{r}_1, \boldsymbol{r}_1') = U^C(\boldsymbol{r}_1, \boldsymbol{r}_1') + U_R(\boldsymbol{r}_1)\delta(\boldsymbol{r}_1 - \boldsymbol{r}_1'). \tag{6.6}$$

Then introducing the definition

$$\varphi_J(\boldsymbol{r}) = \frac{R_{nlJ}(r)}{r} \mathscr{Y}^m_{Jls}(\hat{\boldsymbol{r}}) \tag{6.7}$$

and recalling that H_0 is just the kinetic energy operator, one has

$$\frac{1}{r_1}\left\{\epsilon + \frac{1}{2M}\left(\frac{\mathrm{d}^2}{\mathrm{d}r_1^2} + \frac{l(l+1)}{r_1^2}\right)\right\} R_{nlJ}(r_1)\mathscr{Y}^m_{Jls}(\hat{\boldsymbol{r}}_1)$$
$$= \int \mathrm{d}\boldsymbol{r}_1'\{U'^{(C)}(\boldsymbol{r}_1, \boldsymbol{r}_1') + \boldsymbol{L}_1\cdot\boldsymbol{S}_1 U^{(LS)}(\boldsymbol{r}_1, \boldsymbol{r}_1')\} R_{nlJ}(r_1')\mathscr{Y}^m_{Jls}(\hat{r}_1'). \tag{6.8}$$

Now

$$U^{(a)}(\boldsymbol{r}_1, \boldsymbol{r}_1') = \sum_l (2l+1) U^{(a)}(r_1, r_1') P_l(\hat{\boldsymbol{r}}_1, \hat{\boldsymbol{r}}_1'), \tag{6.9}$$

where $(a) = (LS)$ or $'(C)$. The quantily $\boldsymbol{L}_1\cdot\boldsymbol{S}_1$ is diagonal in $\mathscr{Y}^m_{Jls}$ with eigenvalues

$$\tfrac{1}{2}[J(J+1) - l(l+1) - \tfrac{3}{4}].$$

So that on multiplying on the left by $\mathscr{Y}^m_{Jls}(r_1)^*$ and integrating over the angles of $\boldsymbol{r}_1$ and $\boldsymbol{r}_1'$,

$$\frac{1}{r_1}\left\{\epsilon + \frac{1}{2M}\left(\frac{\mathrm{d}^2}{\mathrm{d}r_1^2} + \frac{l(l+1)}{r_1^2}\right)\right\} R_{nlJ}(r_1)$$
$$= 4\pi \int r_1' \,\mathrm{d}r_1'\, U_{Jl}(r_1, r_1') R_{nlJ}(r_1') \tag{6.10}$$

where

$$U_{Jl} = U_l'^{(C)} + \tfrac{1}{2}[J(J+1) - l(l+1) - \tfrac{3}{4}] U^{(LS)}$$

and the relation

$$\sum \mathscr{Y}^m_{Jls}(\hat{\boldsymbol{r}}_1)^* \mathscr{Y}^m_{Jls}(\hat{\boldsymbol{r}}_1') = \left(\frac{2l+1}{4\pi}\right)^{\frac{1}{2}} \frac{N_{Jl}}{(2J+1)} Y^0_l(\hat{\boldsymbol{r}}_1, \hat{\boldsymbol{r}}_1') \tag{6.11}$$

have been used. N_{Jl} is the number of particles in the shell $Jl_{(n)}$, $(= 2J + 1$ for a completely filled shell).

6a (i). REDUCTION OF THE INTEGRO-DIFFERENTIAL EQUATION TO A DIFFERENTIAL EQUATION (B17)

With obvious abbreviations the radial equation becomes

$$(\epsilon - H_0(i)) \frac{R(r)}{r} = 4\pi \int r' \,\mathrm{d}r'\, U(r, r') R(r'). \tag{6.12}$$

In the limiting case of a local potential one has

$$U(r, r') \to U(r)\delta(r - r')/4\pi r^2$$

and the equation reduces to an ordinary Schrödinger equation,

$$(\epsilon - H_0(i)) R(r) = U(r) R(r). \tag{6.13}$$

A straightforward simplification on (12) would be to use the fact that the range of non-locality of $U(r, r')$ is small and expand $R(r')$ on the right hand side about r,

$$R(r') = R(r) + (r - r')\frac{\mathrm{d}R}{\mathrm{d}r} + \ldots.$$

Then we have a simple differential equation, if higher derivatives are neglected.

$$(\epsilon - H_0(i))R(r) = F(r)R(r) + G(r)\frac{\mathrm{d}R(r)}{\mathrm{d}r}$$

where

$$F(r) = 4\pi \int rr' U(r, r')\,\mathrm{d}r'$$

$$G(r) = 4\pi \int rr'(r - r')U(r, r')\,\mathrm{d}r'.$$

The solution of this equation may be expected to be the same as those of (13) only if the range of non-locality is very small. BGW observed that a redefinition of F and G can be made such that at the convergence of a *successful* iteration procedure the solutions agree with those of Eq. (12). Let the $(n + 1)$th approximation to the solution satisfy the equation

$$(\epsilon - H_0(i))R^{n+1}(r) = F^n(r)R^{n+1}(r) + G^n(r)\frac{\mathrm{d}R^{n+1}}{\mathrm{d}r} \tag{6.14}$$

where

$$F^n(r) = 4\pi \int rr'\,\mathrm{d}r' \frac{U(r, r')}{D^n(r)}\left[R^n(r')R^n(r) + a^2\frac{\mathrm{d}R^n(r')}{\mathrm{d}r'}\frac{\mathrm{d}R^n(r)}{\mathrm{d}r}\right] \tag{6.15a}$$

$$G^n(r) = 4\pi a^2 \int rr'\,\mathrm{d}r' \frac{U(r, r')}{D^n(r)}\left[R^n(r')\frac{\mathrm{d}R^n(r)}{\mathrm{d}r} - R^n(r)\frac{\mathrm{d}R^n(r')}{\mathrm{d}r'}\right] \tag{6.15b}$$

$$D^n(r) = [R^n(r)]^2 + a^2\left[\frac{\mathrm{d}R^n(r)}{\mathrm{d}r}\right]^2 \tag{6.15c}$$

where a is a parameter of dimension of length and is chosen to be of the order of the range of non-locality or the 'healing distance' (0.5 to 1.0 fm). These forms of F^n and G^n ensure that (14) has the important feature of the non-local equation (12) that the right hand side does not vanish at the zeroes of $R(r)$ (cf. r.h.s. of (13)). Also in the limit of locality it goes into the correct local equation (13). Above all at convergence, where $R^{n+1} = R^n$, (14) reduces to the integral equation (12) so that $R^n = R$.

6a (ii). Approximate density dependence of the t-matrix

The contribution to the t-matrix from the attractive part of the potential does not depend very strongly on the density but the contribution from the core does, so that one may write

$$t(\rho) = t_a + t_c(\rho).$$

In the earlier approximate form

$$t_c(\rho) = \frac{A(\rho)}{4\pi r_c^2}\,\delta(r_{12} - r_c)\delta(r'_{12} - r_c)$$

the coefficient A, which has different values for singlet and triplet states, has the form

$$A \simeq a(1 - b/r_0)^{-1}.$$

The constants a and b are evaluated from earlier results on the nuclear matter problem (B19).

In this way U_R was found in ref. (B25) to be

$$U_R \simeq (240\ \mathrm{MeV})\rho^2\ (\mathrm{fm})^6$$

which is quite a significant contribution.

6b. *Eden's method (E4)*

This method exploits the peculiar properties of the Harmonic Oscillator (H.O.) potential. This potential has been successfully used in many shell model calculations and therefore it is very reasonable to base a calculation on it as the model single particle potential. For this case the wave functions of the centre of mass and relative motion separate quite easily in the energy denominators of the t-matrix. If n_1 and n_2 are the total quantum numbers of the motion of the two particles and n and N those of their relative and centre of mass motion respectively, then

$$n_1 + n_2 = n + N \tag{6.16}$$

and the associated energies are

$$\epsilon_{n_1} + \epsilon_{n_2} = (n_1 + n_2 + 3)\hbar\omega = \epsilon_n + \epsilon_N = (n + N + 3)\hbar\omega. \tag{6.17}$$

The main point of Eden's method is the observation that there exists a workable approximation in which the Pauli principle can be taken into account. (Recall the coupling that the Pauli principle introduces between the centre of mass and relative motion.) In the case of

O^{16}, for instance, the states with $n_i = 0$ and 1 are occupied so that the Pauli principle forbids the intermediate states $n_i = 0$ or 1. Eden's approximation consists in omitting states with $n_1 + n_2 = n + N \leqslant 4$ which is believed to take into account the essential features of the Pauli principle effects (see Fig. 33).

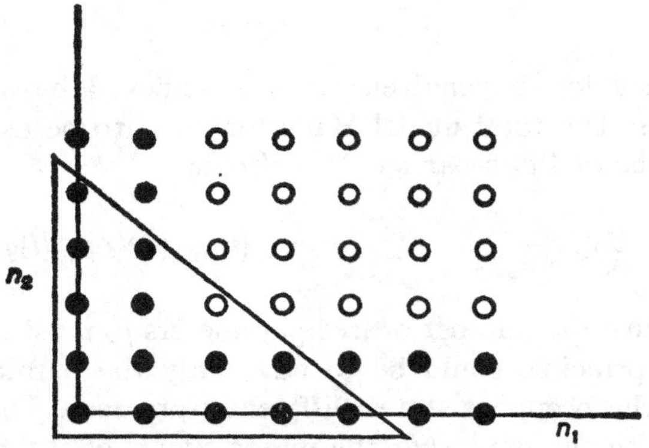

Fig. 33. The Pauli principle in O^{16} forbids all the states denoted by ● for two-particle excitations. Only the states denoted by ○ are allowed. Eden's approximation (E4) consists in forbidding the states inside the triangle instead.

Within these approximations the t-matrix is determined as a function of a single parameter α which determines the strength of the single particle potential. [Certain simplifications in the details of evaluation of the t-matrix are introduced by using a so-called generalised perturbation method (E4). The chief feature of this method is to write the potential as a sum of two parts, one of which, that containing the repulsive core, is treated by t-matrix methods and the other by ordinary perturbation theory.]

The next major approximation of this method is in the treatment of self-consistency problems. In the usual form it is required that

$$(p|U|l) = \sum_{\substack{m \\ \text{(occupied)}}} (pm|t|lm - ml) \tag{6.18}$$

and to achieve this one varies the wave functions. It is noted now that the above relation can be satisfied for diagonal elements by introducing a state dependent potential (this is analogous to the velocity dependent potentials of the nuclear matter case). We modify the model Hamiltonian

$$H_0 \to H_0' = H_0 - \sum_n \mu_n |\varphi_n\rangle \langle\varphi_n|, \tag{6.19}$$

so that the diagonal element of the potential becomes

$$(l|U|l) = \tfrac{1}{2}\epsilon_l - \mu_l. \tag{6.20}$$

Self-consistency for diagonal elements is achieved by adjusting the parameters μ_l. The total model Hamiltonian is to be expressed as a sum of symmetrised two-particle Hamiltonians

$$H_{ij} = H_0(i) + H_0(j) - \sum_{\nu=0}^{\infty} \sum_{m_1+m_2=\nu} \mu_\nu |\varphi_{m_1}(i)\varphi_{m_2}(j)\rangle \langle\varphi_{m_1}(i)\varphi_{m_2}(j)|. \tag{6.21}$$

For simplicity the number of free parameters μ_ν must be restricted. One guiding principle could be to have only the parameters corresponding to the occupied states different from zero. This will be in accordance with the hope that the major effects of state dependence should manifest in the occupied states. For O^{16} Eden sets

$$\mu_\nu = 0; \quad \nu \geqslant 3 \tag{6.22}$$

and further restricts it to a one parameter problem by setting

$$\mu_0 = 2\mu_2; \quad 2\mu_1 = 3\mu_2 \tag{6.22'}$$

which corresponds to an adjustable (1s — 1p) level separation which is to be equal to the (1p — 1d) level separation, while self-consistency effect on 1d levels is neglected.

For each value of the oscillator strength parameter α the condition implied by (18) and (20) is satisfied by varying the μ's. Since the energy depends only on these diagonal elements one may say that this variation makes the eigenvalues approximately self-consistent. Then keeping the t-matrix fixed as a function of α, the energy is minimised w.r.t. α

$$E(\alpha) = \sum_m (m|T|m) + \tfrac{1}{2} \sum_{m,n} (mn|t|mn - nm)$$

$$\left.\frac{\partial E}{\partial \alpha}\right|_{t(\alpha)} = 0. \tag{6.23}$$

The variation here determines α. The dependence on α is brought into the energy by means of the wave functions which are fully determined by α. Hence this minimisation makes the wave functions approximately self-consistent. Such separation in wave function and eigenvalue self-consistency is, of course, artificial. The true requirement is expressed by (18). To show the connection of the variational principle to (18), one observes that the differentiation w.r.t. α is only on the wave functions.

The radial parts of the wave functions are, for instance, ($\beta = \sqrt{2\alpha}$)

$$R_{00} = 2\pi^{-\frac{1}{4}}\beta^{\frac{3}{2}}\, \mathrm{e}^{-\frac{1}{2}\beta^2 r^2} \tag{6.24a}$$

$$R_{11} = \pi^{-\frac{1}{4}}\sqrt{\tfrac{8}{3}}\, \beta^{\frac{5}{2}} r\, \mathrm{e}^{-\frac{1}{2}\beta^2 r^2} \tag{6.24b}$$

and their derivatives are

$$\frac{\partial R_{00}}{\partial \beta} = \sqrt{\tfrac{3}{2}}\, R_{20}, \qquad \frac{\partial R_{11}}{\partial \beta} = \sqrt{\tfrac{5}{2}}\, R_{31}. \tag{6.25}$$

If one writes the states obtained by differentiation as follows

$$|m') = \frac{\partial}{\partial \beta}\, |m) \tag{6.26}$$

then the minimisation condition (23) becomes

$$\sum_{m} (m'|U|m) = \sum_{m,n} (m'n|t|mn - nm) \tag{6.27}$$

where m and n are in the chosen configuration. In deriving the above relation

$$(m'|H_0|m) = 0 = (m'|T|m) + (m'|U|m) \tag{6.28}$$

has been used.

It is seen that the variational principle is equivalent to a self-consistency requirement for some of the non-diagonal elements. It is an approximation to the 'exact' relation (18) in as much as m' are some special states.

The scheme for the present method is given below.

Model Hamiltonian (6.21) $H(\mu_\nu, \alpha) = \sum_{i<j} H(ij)$

$\downarrow$

$t(\mu_\nu, \alpha)$ – with triangular approximation to the exclusion principle, generalised perturbation theory and restrictions on μ_ν.

$\downarrow$

$\epsilon_l(\alpha, \mu_\nu)$, Eq. (6.18), – μ_ν adjusted for eigenvalue self-consistency (Eq. (6.20)).

$\downarrow$

$\alpha_{\min}$ Eq. (6.23) $E(\alpha_{\min})$.

Note that there is no iterative 'loop' here as in the original method or as in the BGW method. Even the usual Hartree-Fock problem is avoided. The present method also suffers from the defect of not having been corrected for the centre of mass motion.

6c. *Centre of mass motion (B10, L2)*

The problem as discussed so far involves establishing a connection between spectra of a real Hamiltonian

$$H = \sum_i T_i + \sum_{i<j} v_{ij}$$

and that of a model Hamiltonian

$$H_0 = \sum_i T_i + \sum_i U_i.$$

As mentioned in the previous section U_i is to be taken as fixed in space. This is also done in the shell model. Then there appears a tremendous qualitative difference between the centre of mass motion in H and H_0. In H there is a continuous spectrum for centre of mass motion whereas in H_0 it is bounded. Thus any treatment starting from H_0 would converge very badly.

One prescription (B10, L2) for avoiding this difficulty is to put the entire original system in a Harmonic Oscillator so that the centre of mass motion is in well known, recognizeable bound states. This will not affect the spectrum of internal motion which is the only interesting spectrum in the problem. Then there will be no difficulty of convergence when the model system is started off in a given state.

To restrict the centre of mass motion one adds a function of centre of mass coordinate to H

$$H' = H + \frac{M\omega^2}{2A} (\sum_i x_i)^2 = \sum_i T_i + \frac{M\omega^2}{2} \sum_i x_i^2 + \sum_{i<j} v'_{ij} \quad (6.29)$$

$$v'_{ij} = v_{ij} - \frac{M\omega^2}{A} (x_i - x_j)^2 \quad (6.30)$$

or

$$H' = H_{\text{h.o.}} + \sum_{i<j} v'_{ij}. \quad (6.31)$$

One can start calculations with $H_{\text{h.o.}} = H_0$. This is expected to be quite successful in view of the success of the H.O. potential in the shell model. It simplifies convergence difficulties. The states due to spurious centre of mass motion will have to be recognized and rejected (E6, L6). The only difference is that the new two-particle interaction is now a long-range interaction because of the $(x_i - x_j)^2$ term and those parts of the theory which depend on v'_{ij} being short-ranged might be affected. This prescription has been successfully used by Mang and Wild (M2) in the three and four body problems.

6d. *Accuracy of the calculations*

The difficulties in making comparison of different works have been mentioned before. The chief 'working principle' of individual studies seems to be to make a number of approximations suggested by the experience with similar numerical work in earlier models. Thus, for instance, Eden (E4) draws upon the earlier shell model works in restricting the μ's as in (22'). Brueckner has by now developed a highly intuitive sense for making approximations on the basis of his long experience with many body calculations. Perhaps because of the enormous complexities of these calculations, the question of accuracy is never adequately discussed. If one recalls the labour and care needed to make adequate Hartree-Fock calculations for atoms (e.g., see p. 359 of ref. C1, also H7), one would perhaps be surprised that the present theories are expected to give significant results for nuclei. However, it might be that we have here a framework in which a calculation of nuclear properties with known limits on accuracy can be developed. This will, of course, depend on a successful evaluation of the role of superconductivity effects (Ch. V, sec. 5) and of higher order terms in the level shift expansion. One of the clearer discussions of the magnitude of the first few terms is given in a work of Levinger and others (L4, see also S16).

7. FURTHER APPLICATIONS

7a. *Some parameters for nuclear matter*

As has been mentioned before, a study of nuclear matter is undertaken because we expect that if such a system were possible it would saturate at the value of density found in the interior of heavy nuclei and at the value of average binding energy of all natural nuclei. This, of course, is an extrapolation suggested by approximate constancy of these two quantities through most of the periodic table. There are two more parameters which one can calculate for nuclear matter and hope to compare with those known from nuclei. These are the *compressibility* and the *symmetry energy*.

The *compressibility*, K, is defined as

$$K = R^2 \left(\frac{\partial^2 E}{\partial R^2} \right) \Bigg|_{R=R_0} = 9\rho^2 \left(\frac{\partial^2 E}{\partial \rho^2} \right) \Bigg|_{\rho=\rho_0} . \tag{7.1}$$

This is quite easily evaluated once the energy E is known as a function of density. The value obtained by Brueckner (B13) is about 170 MeV. This may be compared with the so-called experimental values of 100 to 150 MeV needed to fit the data on isotopic shift (H5, W11). Earlier estimates of compressibility were obtained by fitting masses and from some considerations of nuclear potential energy (F8). However, it is now well known that the masses can be fitted without introducing the compressibility terms (W12 and other references).

The *symmetry energy* is the name given to the coefficient in the term proportional to $(N - Z)^2/A$ in the mass formula (e.g., W12 or some nuclear physics text-book). The basic idea in calculating this quantity has been described for instance by Fermi (F9). In slightly more general terms the energy of a system of N neutrons and Z protons is determined by the Fermi momenta k_{FN} and k_{FP} for the two types of particles.

$$E = E(k_{\mathrm{FN}}, k_{\mathrm{FP}}) \tag{7.2}$$

where

$$k_{\mathrm{FN}} = k_{\mathrm{F}}(1+\varepsilon)^{\frac{1}{3}}; \quad k_{\mathrm{FP}} = k_{\mathrm{F}}(1-\varepsilon)^{\frac{1}{3}} \tag{7.3}$$

and

$$\varepsilon = (N-Z)/(N+Z) \tag{7.4}$$

and k_{F} is the Fermi momentum for $\varepsilon = 0$. Because of charge independence and neglect of the Coulomb forces

$$\lim_{\varepsilon\to 0} E(k_{\mathrm{FN}}, k'_{\mathrm{FP}}) = \lim_{\varepsilon\to 0} E(k'_{\mathrm{FP}}, k_{\mathrm{FN}}). \tag{7.5}$$

In the Taylor expansion of (2) about $\varepsilon = 0$ the first derivative of E vanishes because of saturation. Hence up to second order in ε

$$E(k_{\mathrm{FN}}, k_{\mathrm{FP}}) = E(k_{\mathrm{F}}, k_{\mathrm{F}}) + \tfrac{2}{9}\left(k_{\mathrm{F}}^2 \frac{\partial^2 E}{\partial k_{\mathrm{F}}^2}\right)\Bigg|_{\varepsilon=0} \varepsilon^2. \tag{7.6}$$

The coefficient of ε^2 which may also be expressed in terms of density derivatives is evaluated as in the case of compressibility. The values obtained are in fair agreement with those obtained from the mass formula (W12) although the mass formula values vary significantly depending on the criterion of fit (K12).

In an early paper Brueckner (B27) made some evaluations of the surface and distortion energies for the nuclear matter. These estimates have no direct relevance to the surface and distortions of an actual nucleus. These two effects are closely connected with the finite extension of nuclei and a satisfactory solution amounts almost to solving the problem of the finite nucleus. For the complexities involved even in approximate treatment of these problems, see references (S13, W13) on surface effects and (N1, K13) on deformed nuclei.

7b. *Shell model*

The question of defining the single particle potential has been discussed in detail in Ch. III, sec. 6 and Ch. IV, sec. 4b, 6, and that of calculating the total energy and wave function has been the main concern in all of the foregoing. In this section we bring together these various elements to give a connected account of how the shell model ought to be applied in the light of the present theory. It might be mentioned that we are perhaps very far from a rigorous mathematical justification of the nuclear single particle models, such a justification for the single particle model for *atoms* was published in 1951 by Kato (K10).

The model or zero-order Hamiltonian is defined as

$$H_M = H_0 = \sum_i (T_i + U_i)$$

$$U_i = \sum_{i \neq j} (ij|t|ij - ji).$$

For cases where the shell model is applicable, the self-consistent potential U_i should turn out to be spherically symmetric. Experience with the shell model suggests that the shape of this potential should be similar to that of the Harmonic Oscillator for light nuclei (E4, M2), and of a Saxon-Wood type for heavier nuclei (R3). An important feature of the new self-consistent potential is expected to be its non-locality (sec. 4b; B15, B17). The effect of this non-locality is to some extent equivalent to taking a position dependent effective mass with a local potential. The same effect may be simulated by introducing a state dependent single particle potential (E4). Although this non-local effect in single particle potential is very important as far as questions of self-consistency are concerned, it is not yet clear if it implies any radical difference [48] of principle in application of the shell model (F10, G8, R4, T8). In particular, one can still set up the usual uncorrelated determinantal wave function for the model configurations (G8, R4).

The first significant difference is that the present method shows the true wave function Ψ, given by (I. 2.18), (II. 4.47), etc., to contain strong correlations between particles. Indeed, experiments at high energies confirm the existence of these correlations in nuclei (B28). Correlations are brought about especially because the hard core forces the wave function to vanish whenever the interparticle distance is less than the core radius. The resulting distortion of the wave function gives rise to high momentum components in the wave

[48] Sometimes one defines an equivalent local potential for a non-local one. An obvious definition would be to write $\int U(\mathbf{r}, \mathbf{r}')\varphi_i(\mathbf{r}')\,d\mathbf{r}' = U_{\text{eq}}(r)\varphi_i(r)$. It is evident that the equivalent potential, apart from being strongly state dependent, has in general a peculiar dependence on r, it has singularities at the zeros of φ_i and changes sign in some regions. Brueckner *et al.* (B18, B35) have displayed some equivalent local potentials arising in their work on nuclei. They define $F(r)$ of (6.15a) to be the equivalent local potential.

function. The presence of high momentum components in the language of shell model means large amplitudes for higher configurations in the true wave function Ψ. As a matter of fact, from (II. 4.91) and (II. 4.104) it is seen that the probability of finding the chosen configuration Φ_0 in the corresponding true wave function Ψ_0 decreases as $e^{-\alpha A}$, where A is the number of particles (see also Bethe (B15, equation (6.7)). This implies that the single particle description is not good and of course it *is not* good at high energies. The point to be remembered here is that the shell model works only at low energies and for quantities such as magnetic moments which are expectation values of single particle operators. These quantities may be calculated using the shell model wave function Φ_0 provided consistent use is made of the fact that the true wave function Ψ_0 is given by

$$\Psi_0 = M\Phi_0$$

so that the expectation value of any operator O is

$$(\Psi_0, O\Psi_0) = (M\Phi_0, OM\Phi_0) = (\Phi_0, M^*OM\Phi_0) = (\Phi_0, O_M\Phi_0)$$

$$O_M = M^*OM.$$

The operator O_M is the so-called model operator (B12, E1) and it may be shown that it does not differ much from O for operators which are well represented in the shell model scheme. In particular, the electromagnetic operators have been discussed in this way (B29).

In our discussions we have assumed, for sake of simplicity, that the chosen configuration is non-degenerate. Only the ground state configurations of nuclei with closed-shells, or closed-shells-plus or minus-one nucleon satisfy this condition. Degeneracy sets in for two or more nucleons outside the closed shell and for excited configurations in general. This situation is similar to the atomic case for which the perturbation theory can be set up in the well known way (C1). In the nuclear case the degeneracy is very high and the interactions are singular. The procedure for treating such cases was given by Bethe (B15). We give here a similar discussion which is adapted to the notation and language of this study.

When there are a number of states degenerate with the chosen configuration, two cases arise: one, as mentioned above, where degenerate configurations perturb each other's energy but no real transitions take place (stationary states), and the other where real transitions allowed by energy conservation take place between degenerate levels (metastable states). The two cases correspond to different choices of boundary conditions for the wave functions or to different ways of treating the singularity of the propagator $1/(E - H_0)$ in the perturbation series.

It is easier to discuss the *stationary states* from the point of view of Ch. III, where the t-matrix equation

$$t = v + v\frac{1}{a}t$$

is just an auxiliary for solving the actual Schrödinger equation of the system. We can define the propagator in the above equation such that it becomes zero for all states degenerate with the chosen configuration Φ_C. The t-matrix obtained by solving this equation will include the effect of all configurations not degenerate with Φ_C. For an excited state Φ_C there will be intermediate states of lower energy so that the propagator will change sign, but this causes no difficulty. The important difference arises when v is eliminated in terms of t in the Schrödinger equation. One must now diagonalise V among the degenerate states. To the first order in t then one must diagonalise the matrix

$$(\Phi_B|\Sigma\, T_i + \sum_{i<j} t_{ij}|\Phi_C),$$

where Φ_B are configurations degenerate with Φ_C. Actually, of course, one must use the whole series (III. 2.2) instead of just t_{ij}. We expect the potential energy to be finite because we are already including the effect of higher configurations and building in the required correlations.

The usual procedure of shell model in which an effective nonsingular potential is diagonalised among the degenerate states for finding energy levels, is thus justified. Indeed we see here how the effective potential, compatible with two nucleon potentials, may be calculated. Also it is evident that mixtures of higher configurations in Φ_C should not be allowed since they are already contained in the t-matrix [49].

[49] This is true only to first or second order in t-matrix .The actual situation will be rendered more complex by considerations similar to those following (III. 5.1). Empirical evidence for configuration dependence of the effective two body interaction has accumulated over the years. One of the clearest cases is discussed by Pandya and Shah (P4, see also B40 and P5).

The problem of diagonalising the entire Hamiltonian is thus split into two simpler problems, viz., that of calculating the effective potential and that of diagonalising a finite matrix [50] in degenerate configurations. The flexibility in defining the propagator in t-matrix equations allows us a certain freedom in choosing the states we take to be degenerate. This freedom may be used to ensure best convergence for a given amount of labour.

Treatment of metastable states – the optical model problem – is complicated by considerations mentioned in Ch. II, secs. 4f (ii) and (iii). A simple treatment related to the present problem is as follows (K15, S8):

Consider a matrix $t^{(+)}$ defined by [51]

$$t^{(+)} = v + \lim_{\alpha \to 0} v(E_0 - H_0 + i\alpha)^{-1} t^{(+)}$$

and the t-matrix defined by

$$t = v + v \frac{P}{E_0 - H_0} t.$$

Since

$$\lim_{\alpha \to 0} \frac{1}{E_0 - H_0 + i\alpha} = \frac{P}{E_0 - H_0} - i\pi\delta(E_0 - H_0)$$

we have

$$t^{(+)} = v + v \frac{P}{E_0 - H_0} t^{(+)} - i\pi v \delta(E_0 - H_0) t^{(+)}$$

or

$$\left(1 - v \frac{P}{E_0 - H_0}\right) (t^{(+)} - t + i\pi t \delta(E_0 - H_0) t^{(+)}) = 0.$$

[50] This matrix rapidly increases in dimensions as the excitation or the number of particles outside the shell increases. The shell model breaks down in such cases and one has to go over to a collective description (A3, B30). An attempt to describe collective motions in terms of the t-matrix has been made by Glassgold *et al.* (G7). Brueckner and Thieberger have considered dipole oscillations of nuclear matter (B31) in similar terms.

[51] An ambiguity in this limiting process which may sometimes lead to errors has been discussed by Foldy and Tobocman (F14).

Hence for those states for which

$$1 - v \frac{P}{E_0 - H_0} \neq 0$$

$$t^{(+)} = t + i\pi t \delta(E_0 - H_0) t^{(+)}.$$

Since t is real it follows that $t^{(+)}$ must be complex. In first approximation the real part of $t^{(+)}$ is just t. The operator $t^{(+)}$ describes the effective interaction experienced by an outgoing particle inside the nucleus. The appropriate single particle potential may be calculated from $t^{(+)}$ using the methods of section 4b. The imaginary part of this potential (G9, J3, S8) is responsible for absorption of the particle in the nucleus and the real part is basically the same as the shell model potential for a nucleon.

LIST OF IMPORTANT SYMBOLS IN CHAPTER IV

(N.B. *Only new symbols or those with different meanings are listed here; the rest are as in Chs. II and III*)

a	constant of non-locality in BGW equation (6.15)
a_k	scattering amplitude for particles with relative momentum $\boldsymbol{k}$
$A(\boldsymbol{r}, \boldsymbol{r}')$	spin independent part of the non-local matrix
$B(\boldsymbol{r}, \boldsymbol{r}')$	coefficient of the spin orbit part of the non-local t-matrix, also $B'(\boldsymbol{r}, \boldsymbol{r}')$, (4.21)
$C(\boldsymbol{r}, \boldsymbol{r}')$	a part of the t-matrix, (4.21)
C_l	a constant occurring in the partial wave expansion, (3.1)
$D(\boldsymbol{r}, \boldsymbol{r}')$	a part of the t-matrix (4.21)
$D(r)$	a function occurring in the BGW equation, (6.15)
$E(= \mathscr{E})$	the self-consistent total energy
$f(P, p)$	cut-off function to take account of the Pauli principle in the two-particle Green's function, (3.20), (3.21), for interacting particles

with total and relative momenta P and p respectively

$f'(P, p)$ an approximation to $f(P, p)$, (3.22)

$F(r)$ effective single particle potential in BGW method, (6.15)

$F^{(n)}(r)$ n^{th} iterate for $F(r)$, (6.15)

$F(r, r')$ effective Green's function after approximate treatment of the hard-core, (3.29)

$F_l(r, r')$ $F(r, r')$ corresponding to l^{th} partial wave

G generic symbol for Green's function

$G_l(r, r')$ part of the Green's function corresponding to l^{th} partial wave

$G(r)$ a function in BGW equation, (6.15)

$G^{(n)}(r)$ n^{th} iterate of $G(r)$

$H_0(i)$ unperturbed single particle Hamiltonian,

H_{ij} symmetrised two-particle perturbed Hamiltonian, (6.21)

H' total Hamiltonian modified to restrict centre of mass motion, (6.29)

H_M model Hamiltonian

$\mathscr{H}$ model two-particle Hamiltonian, (1.16)

K compressibility, (7.1)

M when not used as an operator – mass of the particle (nucleon)

N number of neutrons in the nucleus

N_n number of particles in the level n

N_{Jl} number of particles in the same Jl level of a single particle potential

$\boldsymbol{P}$ total centre of mass momentum of two particles

$Q(\boldsymbol{r}, \boldsymbol{r}')$ projection operator to take account of the Pauli principle (1.17)

$R_{nlJ}(r)$ single particle radial wave functions

$S(r)$ zero order wave function in Bethe-Goldstone equation after the effect of the hard-core has been included, (3.29)

$S_l(r)$, $S(r)$ for l^{th} partial wave

$t(\delta E)$ approximate t-matrix for off-the-energy-shell propagation, (1.3)

$t(\delta E, P)$ $t(\delta E)$ when the two particles have a total momentum P inside the nucleus, (1.18)

$t^{(+)}$ effective two-particle interaction for outgoing particles

t_a part of the t-interaction, sec. 6a (ii)

$t_c(\rho)$ density dependent part of the t-matrix arising

	from the hard-core, sec. 6a (ii)
U	generic symbol for single particle potential
U_R	rearrangement potential
$\mathscr{U}(r)$	radial part of the relative two-particle wave function. ($\mathscr{U}_0(r)$ – unperturbed, $\mathscr{U}(r)$ – perturbed)
$\mathscr{U}^{Js}_{ll'}(r)$	$\mathscr{U}(r)$ with indices fully specified
v_a, v_{core}, v_C, v_T, v_{LS}	attractive, core, central, tensor and spin orbit parts of the two-particle interaction
$v^{Js}_{ll'}(r)$	matrix element of v for angular parts, (3.9)
$w(\boldsymbol{p}, \boldsymbol{p}')$	momentum space representation of a general (non-local) two-particle potential (2.6)
$\tilde{w}(\boldsymbol{p})$	a factor in the separable non-local two-particle interaction
$\mathscr{Y}^M_{Jls}$	the angular momentum part of the two-particle wave function, (3.2)
α	oscillator strength parameter
β	oscillator strength parameter ($= \sqrt{2}\alpha$)
Δ	energy connected to rearrangement potential, (5.1), (5.7)
δ_l	phase shifts
λ, $\lambda^{Js}_{ll'}$	constants used to take into account the effect of hard-core, (3.26)
Λ_t, Λ_s	triplet and singlet spin projection operators, (4.37)
ϵ_p	single particle energy of a particle with momentum p or the energy of relative motion of two particles with relative momentum p
Φ_{12}, $\Phi_{l_1l_2}(\boldsymbol{r}_1\boldsymbol{r}_2)$; $\Phi_{p,P}$,	various forms of unperturbed two-particle wave functions
Φ_p	relative part of the unperturbed two-particle wave functions
Ψ_{12}, $\Psi_{l_1l_2}(\boldsymbol{r}_1\boldsymbol{r}_2)$; $\Psi_{p,P}$,	various forms of the perturbed two-particle wave function
Ψ_p	relative part of the perturbed two-particle wave function
Ω_{12}	the wave matrix – same as M_{12}

CHAPTER V

CONNECTION WITH SOME OTHER METHODS

Some other methods of handling the many body problem will now be discussed to show their relationship to the Brueckner method and to perturbation theory. Here also the implicit assumption is the weakness of the perturbing effects, although the exact criterion of weakness will be difficult to establish. The connections are not fully worked out in any of the following cases. It might not even be possible to do so in a mathematically satisfactory manner for most of them. But a certain intuitive insight may be gained here by comparing the final results.

1. JASTROW'S VARIATIONAL METHOD (A2, C3, D5, I1, J2, T4)

In this method the difficulty associated with the hard cores is sought to be overcome by taking a trial wave function of the form

$$\Psi = [\prod_{i<j} (1 + f(r_{ij})]\Phi \tag{1.1}$$

where Φ is the usual independent particle determinantal wave function. The functions $f(r_{ij})$ are such that they vanish inside the core, (where the potential to begin with may be assumed to be very large but finite, so that no $0 \times \infty$ type situation occurs) and rapidly become equal to one outside the core. A popular form to assume is (e.g. T4)

$$\begin{aligned} [1 + f(r_{ij})] &= 0 \qquad r \leqslant r_c \\ &= 1 - \frac{r_c}{r} e^{-\mu(r-r_c)} \qquad r > r_c. \end{aligned} \tag{1.2}$$

The resulting energy expression is quite complicated. 'Cluster integral' methods for writing down the full energy expression have been devised by Iwamoto and Yamada (I1) and further refinements considered by Clark and Feenberg (C3) and by Fujita (F12). These clusters have no direct relation to the linked clusters of the Brueckner method. The energy obtained is minimised with respect to the parameters in $f(r_{ij})$. These values of the parameters determine the correlation distance of particles. If the core radius does not depend on the state of interacting particles then the simple, state independent, form of $f(r_{ij})$ can be expected to work. If, however, the core radius depends on the state of the particles, as it does in modern potentials given by Gammel and Thaler and others, then this simple form of $f(r_{ij})$ is obviously inadequate. The situation can be remedied quite easily by writing (T4, W7)

$$\Psi = \frac{1}{A!} \sum_{P_l} (-)^{P_l} P_l [\underbrace{\varphi_{l_1}(1)\varphi_{l_2}(2) \ldots \varphi_{l_A}(A)}_{\text{all particles}}$$
$$\times \underbrace{\{(1 + f_{l_1 l_2}(12)) \ldots (1 + f_{l_{A-1} l_A}(A-1, A))\}}_{\text{all pairs}}] \tag{1.3}$$

where the correlation function is now state dependent and P_l is the permutation operator acting on the indices l. This seemingly minor modification enormously complicates actual calculations.

Formally, such a procedure always gives finite results. There are no questions of convergence as in perturbation theory because the number of terms, although large, is always finite. The difficulties arise at the practical level because the finite energy series of the cluster development has to be broken off at some step and only this part of the energy is minimised. One has to ensure that the minimum so obtained is not affected by higher, neglected terms. In this sense one is perhaps as badly off as in perturbation theory or Brueckner theory.

Comparison with Brueckner theory is made through the wave functions of the two theories. Weisskopf and De Shalit (W7) have shown that a wave function of the above form satisfies the many body Schrödinger equation if the terms arising from the close presence of three or more particles can be neglected and if the state dependent correlation function is obtained from the unantisymmetrised solutions of the Bethe-Goldstone equation by means of the following relation:

$$f_{l_1 l_2}(ij) = \frac{\Psi_{l_1 l_2}(ij) - \varphi_{l_1}(i)\varphi_{l_2}(j)}{\varphi_{l_1}(i)\varphi_{l_2}(j)} \tag{1.4}$$

where $\Psi_{l_1 l_2}(ij)$ is the solution of the Bethe-Goldstone equation and φ_{l_1}, φ_{l_2} are single particle model wave functions. As the distance between the two particles becomes large, $f_{l_1 l_2}$ goes to zero since

$$\Psi_{l_1 l_2}(1, 2) \to \varphi_{l_1}(1)\varphi_{l_2}(2) \quad \text{as} \quad |r_{12}| \to \infty. \tag{1.5}$$

However, $f_{l_1 l_2}$ of (4) has poles where either φ_{l_1} or φ_{l_2} have zeros. This makes it look rather different from the f given by (2) but integrated effects are similar.

A slightly different form of this variational principle, based on the idea of separating the nuclear interaction in short and long range parts, has been discussed by Austern and Iano (A4).

2. METHOD OF SUPERPOSITION OF CONFIGURATIONS

Ultimately the wave function obtained from Brueckner theory is also written as a linear combination of determinantal wave functions. This is a superposition of configurations which the theory gives us automatically. Following Nesbet (N2, see also Kromhout K6) one can write the theory in such a way that the superposition aspect is more clearly manifested. Recall the Hamiltonian

$$H = \sum_i T_i + \sum_{i<j} v_{ij} = H_0 + \sum_{i<j} u_{ij}; \quad H\Psi_0 = \mathscr{E}\Psi_0. \tag{2.1}$$

The solution Ψ_0 can be expressed as a linear combination

$$\Psi_0 = \sum_\mu C_{\mu 0}\Phi_\mu; \quad H_0\Phi_\mu = E_\mu\Phi_\mu. \tag{2.2}$$

From Eq. (I. 2.18) $C_{\mu 0}$ are the matrix elements of the M-matrix ($K_{\text{op}} \equiv H_0$, $V = \sum_{i<j} \bar{v}_{ij}$).

$$C_{\mu 0} = \langle \mu | M | 0 \rangle. \tag{2.3}$$

We have, from the wave equation

$$C_{\mu 0} = \langle \mu | M | 0 \rangle = \sum_\nu \frac{N_{\mu\nu}}{\mathscr{E} - D_\nu} C_{\nu 0}; \quad C_{00} = 1 \tag{2.4}$$

where

$$D_{\mu\nu} = \delta_{\mu\nu} \langle \mu | H | \mu \rangle = \delta_{\mu\nu} D_\mu, \quad N_{\mu\nu} = (1 - \delta_{\mu\nu}) \langle \mu | H | \nu \rangle. \tag{2.5}$$

A new operator defined by

$$G = (\mathscr{E} - D)M \tag{2.6}$$

has matrix elements

$$G_{\mu 0} = \sum_\nu \frac{N_{\mu\nu} G_{\nu 0}}{\mathscr{E} - D_\nu}. \tag{2.7}$$

Since we have identically

$$\mathscr{E} = D_0 + G_{00} \tag{2.8}$$

we can write

$$G_{\mu 0} = N_{\mu 0} + \sum_{\mu \neq \nu} \frac{N_{\mu\nu} G_{\nu 0}}{\mathscr{E} - D_\nu}. \tag{2.9}$$

A natural classification of the matrix elements of G or M can be given in terms of the nature of the configurations Φ_μ in relation to the chosen configuration Φ_0. The matrix elements of H (or N and D) between any two determinantal wave functions made from orthonormal single particle states which differ in more than two excitations, vanish and this is the basis of the classification. Let the determinant Φ_0 be made out of the states $m_1, m_2, m_3 \ldots < N$, where N is the number of particles in the system and let $k_1, k_2, k_3 \ldots \geqslant N$ be the other available single particle states. The complete set of Φ_μ may be divided into the following types

$$\Phi_0,\ \Phi^{k_1}_{m_1},\ \Phi^{k_1 k_2}_{m_1 m_2},\ \Phi^{k_1 k_2 k_3}_{m_1 m_2 m_3},\ \ldots$$

where $\Phi^{k_1}_{m_1}$ is formed from Φ_0 by replacing everywhere the orbital φ_{m_1} by the orbital φ_{k_1}, $\Phi^{k_1 k_2}_{m_1 m_2}$ is obtained from $\Phi^{k_1}_{m_1}$ by replacing φ_{m_2} by φ_{k_2} and so on for higher configuration.

The first few members of the infinite set of coupled integral equations that determine the wave function are

$$[\mathscr{E} - D_0] \langle 0|M|0\rangle = \sum_{m_1 k_1} \langle 0|H|k_1; m_1\rangle \langle m_1; k_1|M|0\rangle$$

$$+ \sum_{m_1 m_2 k_1 k_2} \langle 0|H|k_1 k_2; m_1 m_2\rangle \langle m_2 m_1; k_2 k_1|M|0\rangle \tag{2.10a}$$

$$[\mathscr{E} - D^{k_1'}_{m_1'}] \langle m_1'; k_1'|M|0\rangle = \langle m_1' k_1'|H|0\rangle$$

$$+ \sum_{m_1 k_1} \langle m_1' k_1'|H|k_1 m_1\rangle \langle m_1 k_1|M|0\rangle$$

$$+ \sum_{m_1 m_2 k_1 k_2} \langle m_1' k_1'|H|k_1 k_2; m_1 m_2\rangle \langle m_2 m_1; k_2 k_1|M|0\rangle$$

$$+ \text{one more term.} \tag{2.10b}$$

$$[\mathscr{E} - D^{k_1'k_2'}_{m_1'm_2'}] \langle m_1'm_2'; k_1'k_2'|M|0\rangle = \langle m_1'm_2'; k_1'k_2'|H|0\rangle$$

$$+ \sum_{m_1k_1} \langle m_1'm_2'; k_1'k_2'|H|k_1m_1\rangle \langle m_1k_1|M|0\rangle$$

$$+ \sum_{m_1m_2k_1k_2} \langle m_1'm_2'; k_1'k_2'|H|k_1k_2; m_1m_2\rangle \langle m_2m_1; k_2k_1|M|0\rangle$$

$$+ \text{two more terms.} \tag{2.10c}$$

Corresponding equations for the energy or for the G-operator are given by Nesbet (N2). These equations are, of course, exact if a complete set is taken. In practice, however, one needs to work with a truncated basis. From equations (10) one sees that for a determination of energy alone the coefficients $\langle m_1m_2m_3, k_1k_2k_3|M|0\rangle$ are not directly needed. These and higher coefficients influence the energy only indirectly through their influence on the second and first order matrix elements. Thus a reasonable set will be

$$\Phi_0,\ \Phi^{k_1}_{m_1},\ \Phi^{k_1k_2}_{m_1m_2}.$$

In Eqs. (10) all terms arising from this basis are exhibited.

In the above equations, depending on the configurations we wish to include M (or G) may be looked upon as a symmetrised sum of one, two, three-body operators. One can now impose conditions on these operators and the basic vectors Φ_μ to obtain a convergent solution and, in particular, to avoid the difficulties associated with the hard cores.

The Hartree-Fock method imposes a condition on the basis itself that Φ_0 be the best single determinant obtained by minimising the first order energy $\langle 0|H|0\rangle$. This is equivalent to the condition

$$\langle m_1, k_1|H|0\rangle = 0. \tag{2.11}$$

This can be substituted in the previous equation and integral equations obtained for the matrix elements M (or G). The energy, given by Eqs. (10), is

$$\mathscr{E} = \sum_m \langle m|H_0|m\rangle + \sum_{m_1m_2} \langle m_1m_2|\bar{v}|m_1m_2\rangle$$

$$+ \sum_{m_1m_2k_1k_2} \frac{\langle m_1m_2|\bar{v}|k_1k_2\rangle \langle k_1k_2|\bar{G}|m_1m_2\rangle}{\mathscr{E} - D^{k_1k_2}_{m_1m_2}}, \tag{2.12}$$

where $\bar{G}$ is defined by

$$G = (\mathscr{E} - D)M = \sum_{i<j} \bar{G}(ij). \tag{2.13}$$

The integral equation for determining the operator $\bar{G}$ is obtained by substituting in equations (10). If the solution of these equations is used above, then the energy given by (12) would be better than the usual Hartree-Fock result. This question will not be pursued any further.

Brueckner type of integral equation may be obtained by requiring

$$\langle m_1; k_1|M|0\rangle = \langle m_1; k_1|G|0\rangle = 0. \tag{2.14}$$

Under this requirement

$$\mathscr{E} = \langle 0|H|0\rangle + \sum_{m_1m_2k_1k_2} \frac{\langle 0|H|k_1k_2m_1m_2\rangle\,\langle m_2m_1k_2k_1|G|0\rangle}{\mathscr{E} - D^{k_1k_2}_{m_1m_2}} \tag{2.15a}$$

$$0 = \langle m_1'; k_1'|H|0\rangle + \sum \frac{\langle m_1'k_1'|H|k_1k_2m_1m_2\rangle\,\langle m_2m_1k_2k_1|G|0\rangle}{\mathscr{E} - D^{k_1k_2}_{m_1m_2}} \tag{2.15b}$$

$$\langle m_1'm_2'; k_1'k_2'|G|0\rangle = \langle m_1'm_2'k_1'k_2'|H|0\rangle + \sum \frac{\langle m_1'm_2'k_1'k_2'|H|k_1k_2m_1m_2\rangle\,\langle m_2m_1k_2k_1|G|0\rangle}{\mathscr{E} - D^{k_1k_2}_{m_1m_2}}. \tag{2.15c}$$

The last equation for G does not give the corresponding Brueckner t-matrix equation if we make the substitution

$$G = \sum_{i<j} t(ij). \tag{2.16}$$

However, a very similar equation results, viz.,

$$\langle k_1'k_2'|t|m_1'm_2'\rangle = \langle k_1'k_2'|\bar{v}|m_1'm_2'\rangle + \sum_{\substack{k_1k_2\\ \neq k_1'k_2'}} \frac{\langle k_1'k_2'|\bar{v}|k_1k_2\rangle\,\langle k_1k_2|t|m_1'm_2'\rangle}{\mathscr{E} - D^{k_1k_2}_{m_1'm_2'}} + \sum_{\substack{m_1m_2\\ \neq m_1'm_2'}} \frac{\langle k_1'k_2'|t|m_1m_2\rangle\,\langle m_1m_2|\bar{v}|m_1'm_2'\rangle}{\mathscr{E} - D^{k_1'k_2'}_{m_1m_2}}. \tag{2.17}$$

The Pauli principle restriction is contained in the definition of m's and k's.

The last term is equivalent to including hole-hole interaction (C2). Non-linearity now occurs through $\mathscr{E}$, since it is defined in terms of G by (8).

The condition (14) or Eq. (15b) may be looked upon as defining a single particle operator whose eigenfunctions should be used as the basis. This single particle operator is approximately, from (15b),

$$\langle m_1, k_1 | H_0 + \Sigma \left(\bar{v} + \bar{v} \frac{Q}{E - D} \right) t | 0 \rangle$$
$$= \langle m_1, k_1 | H_0 + \sum_{i<j} t(ij) | 0 \rangle. \qquad (2.18)$$

Here Q is an operator which enforces appropriate restrictions on the intermediate states. In the same way the energy expression becomes

$$\mathscr{E} = \langle 0 | H_0 + \sum_{i<j} t(ij) | 0 \rangle. \qquad (2.19)$$

Nesbet (N2) makes some interesting remarks about the relative merits of these so-called Brueckner and Hartree-Fock conditions. But the question of the effect of higher terms seems as intractable here as elsewhere. Perhaps the most interesting observation is that in this method it does not seem quite possible to consider configurations like $\Phi^{k_1 k_2 k_3}_{m_1 m_2 m_3}$ without introducing effective three body potentials

$$\langle m_1 m_2 m_3 ; k_1 k_2 k_3 | G | 0 \rangle \text{ etc.}$$

3. BRENIG'S TWO-PARTICLE APPROXIMATION

This method (B8) [52] is closely related to that of the previous section. Here again one shows how the Bethe-Goldstone equation may be obtained by making approximations on a series of coupled integro-differential equations. The approximation procedure is based on an analogy with the superposition approximation in the theory of liquids. There is a distinct improvement here in as much as it shows that the Bethe-Goldstone equation is not really limited to summing only doubly excited configurations.

[52] A similar method, in density-matrix formalism, has been discussed by Coester and Kümmel (C4).

Let Π be the product of the main diagonal of the determinant $\mathscr{D}_0$,

$$\mathscr{D}_0 = (N!)^{\frac{1}{2}}\Phi_0 \tag{3.1}$$

$$\Pi = \prod_{i=1}^{N} \varphi^i(r_i). \tag{3.2}$$

Then Π^i, Π^{ij}, Π^{ijk} ... are defined by omitting from the product $\prod$ the single particle wave functions $\varphi^i(r_i)$, $\varphi^i(r_i)\varphi^j(r_j)$ ... respectively. *States i, j, k are inside the chosen configuration. These indices are used as abbreviations for m_i, m_j, m_k and also as particle indices.*

Finally, define the correlated wave functions by equations

$$\varphi_i = \int \Pi^i \Psi_0 \, \mathrm{d}\tau' \tag{3.3a}$$

$$\psi_{ij} = \int \Pi^{ij} \Psi_0 \, \mathrm{d}\tau' \tag{3.3b}$$

$$\psi_{ijk} = \int \Pi^{ijk} \Psi_0 \, \mathrm{d}\tau' \tag{3.3c}$$

where integration is over the coordinates present in the Π functions. One can deduce several of their properties straight away. Thus [53]

$$\int \varphi^i(1)^* \varphi_j(1) \, \mathrm{d}1 = \delta_{ij} \tag{3.4a}$$

$$\int \varphi^k(2)^* \psi_{ij}(1, 2) \, \mathrm{d}2 = \varphi_i(1)\delta_{kj} - \varphi_j(1)\delta_{ik} \tag{3.4b}$$

$$\psi_{ij}(1, 2) = -\psi_{ji}(1, 2) = -\psi_{ij}(2, 1). \tag{3.4c}$$

The Schrödinger equation of the many body problem (see (2.1)),

$$H\Psi_0 = \mathscr{E}\Psi_0 \tag{3.5}$$

can be reduced to the following set of coupled integro-differential equations by successively multiplying by various Π-functions and integrating over their coordinates as in (3).

$$\mathscr{E} = (\Pi|H|\Psi_0) = \sum_{1}^{N} (\varphi^i|T_i|\varphi_i) + \tfrac{1}{2} \sum_{i,j}^{N} (\varphi^i\varphi^j|v|\psi_{ij}) \tag{3.6a}$$

$$\mathscr{E}\varphi_i = T_i\varphi_i + \sum_{i} (\varphi^i|v_{ij}|\psi_{ij}) + \sum_{j} (\varphi^j|T_j|\psi_{ij}) + \tfrac{1}{2} \sum_{j,k} (\varphi^j\varphi^k|v|\psi_{ijk}) \tag{3.6b}$$

$$\mathscr{E}\psi_{ij} = (T_i + T_j + v_{ij})\psi_{ij} + \sum_{k} (\varphi^k|v_{ik} + v_{jk}|\psi_{ijk}) + \sum_{k} (\varphi^k|T_k|\psi_{ijk}) + \tfrac{1}{2} \sum_{k,l} (\varphi^k\varphi^l|v_{kl}|\psi_{ijkl}). \tag{3.6c}$$

[53] All coordinates of a particle are represented by numbers 1, 2 ...; d1 means summation and integration over all coordinates of particle 1.

After some algebra and noting relations like

$$\varphi_{m_i}(1) = \varphi^{m_i}(1) + \sum_{k_i} (m_i, k_i|M|0)\varphi^{k_i}(1) \tag{3.7a}$$

$$\psi_{m_i m_j}(12) = \varphi^{m_i m_j}(12) + \sum_{k_i} (m_i, k_i|M|0)\varphi^{k_i m_j}(12) + \sum_{k_j} (m_j, k_j|M|0)\varphi^{m_i k_j}(12) + \sum_{k_i k_j} (m_i m_j k_i k_j|M|0)\varphi^{k_i k_j}(12) \tag{3.7b}$$

$$\varphi^{k_i k_j}(12) = \varphi^{k_i}(1)\varphi^{k_j}(2) - \varphi^{k_i}(2)\varphi^{k_j}(1) \tag{3.7c}$$

one can show that the set (6) is equivalent to set (2.10). This equivalence has been used to calculate the corrections to the Brenig approximation (K8).

The third and fourth terms in (6b) contain the integrals

$$\int \varphi^j(2)^* T_2 \psi_{ij}(1, 2)\, \mathrm{d}2$$

and

$$\int \varphi^j(2)^* \varphi^k(3)^* v_{23} \psi_{ijk}(1, 2, 3)\, \mathrm{d}2\, \mathrm{d}3.$$

In both these integrals 1 is not to be integrated over and its correlation with other particles is unimportant. However, in the second integral it is necessary to retain the full correlations between particles 2 and 3, taking into account the indistinguishability and Pauli principle. The 'two-particle approximation' to (6b) consists in putting in the above integrals

$$\psi_{ij}(12) \cong \varphi_{ij}(12) = \varphi_i(1)\varphi_j(2) - \varphi_i(2)\varphi_j(1) \tag{3.8a}$$

$$\psi_{ijk}(123) \cong \varphi_i(1)\psi_{jk}(23) - \varphi_j(1)\psi_{ik}(23) - \varphi_k(1)\psi_{ji}(23). \tag{3.8b}$$

By choosing φ^i such that (cf. Eq. (2.14), the Brueckner condition):

$$\varphi^i = \varphi_i \tag{3.8c}$$

(6b) finally reduces to the single particle equation

$$(T + U)\varphi^i = \epsilon_i \varphi^i, \tag{3.9a}$$

where

$$\epsilon_i = (\varphi^i|T + U|\varphi^i) = (\varphi^i|T|\varphi^i) + \sum_j (\varphi^i \varphi^j|v|\psi_{ij}). \tag{3.9b}$$

In order to make the two-particle approximation we should remember that here we wish to calculate the two-particle correlation function ψ_{ij}, hence the correlation between particles in the states i and j must be fully preserved. Let

$$\psi_{ij}(12) = \Omega_{12}\varphi_{ij}(12). \tag{3.10}$$

Then for the integral

$$\int \varphi^k(3)^* v_{13}\psi_{ijk}(1, 2, 3)\,\mathrm{d}3$$

of (6c) the two-particle approximation consists in putting

$$\psi_{ijk}(123) \cong \Omega_{13}\{\psi_{ij}(12)\varphi_k(3) - \psi_{ij}(32)\varphi_k(1)\} - \psi_{ij}(13)\varphi_k(2). \tag{3.11}$$

Also, we have

$$U_1\psi_{ij}(12) = \sum_k \int \varphi^k(3)^* v_{13}\Omega_{13}\{\psi_{ij}(12)\varphi_k(3) - \psi_{ij}(32)\varphi_k(1)\}\,\mathrm{d}3 \tag{3.12a}$$

so that

$$\sum_k (\varphi^k(3)\,|v_{13}|\,\psi_{ijk}(123)) = U_1\psi_{ij}(12) - \sum_{k<N} (\varphi^l\varphi^k\,|v|\,\psi_{ij})\varphi_l(1)\varphi_k(2). \tag{3.12b}$$

The approximation for the last term of (6c) is

$$\psi_{ijkl}(1234) \cong \psi_{ij}(12)\psi_{kl}(34) + \text{antisymmetric permutations of } ijkl. \tag{3.13}$$

After some algebra (6c) in this approximation becomes

$$(\mathscr{H}_1 + \mathscr{H}_2 + (1 - \tilde{Q}_{12})v_{12})\psi_{ij}(12) = \epsilon_{ij}\psi_{ij}(12) \tag{3.14}$$

$$\mathscr{H}_1 = T_1 + U_1; \quad \epsilon_{ij} = \epsilon_i + \epsilon_j \tag{3.15}$$

$$\tilde{Q} = \tfrac{1}{2}\sum_{\substack{k<N \\ (l \text{ unrestricted})}} |\varphi_{kl})(\varphi^{kl}| - \tfrac{1}{4}\sum_{k,l<N} |\psi_{kl})(\varphi^{kl}|. \tag{3.16}$$

The operator $(1 - \tilde{Q})$ is a projection operator which introduces the effects of the Pauli principle. The above derivation does not suggest a clear cut procedure for discussing the validity of this approximation or for calculating the corrections. Its merits are simplicity and the fact that it does not involve any expansions in terms of the interaction.

A somewhat better understanding of this approximation is obtained by exploiting the equivalence of the set of equations (6) to

(2.10) and the expansion of correlation functions ψ_{ij}, ψ_{ijk}, etc., given in equations (7). Although one has to deal again with some non-linear integral equations, it can be shown (K8) that if the major effects are well represented in this approximation, then the corrections to the total energy are similar in form to the third (and fourth) order cluster contributions discussed by Bethe (B15) for the Brueckner theory. With the important difference that instead of the t-matrices, which take into account only two-particle correlations, we have three and four particle correlation amplitudes and there are no higher order corrections. It may be supposed that even though the correlation amplitudes in question are, strictly speaking, to be obtained by solving higher order equations in the set (6), they will not bring in any upsetting magnitudes and the Brueckner method may be better than it appears. It opens up a further possibility, viz., that if we can obtain an estimate of three and four particle correlations by some other methods (e.g., the variational method akin to Jastrow's for two-particle correlations), then we shall have another means of checking the validity of this approximation.

4. PSEUDO-POTENTIAL METHOD

This method was first introduced by Fermi (B21, F11) in connection with the problem of scattering of slow neutrons by protons chemically bound in large molecules. The method consists in replacing a boundary condition on the wave function by a (pseudo-)potential introduced in the wave equation. It is thus immediately applicable to a pure hard core potential whose effect is fully taken into account by requiring the wave functions to vanish inside the core. The same idea has been applied to the case of many particles, both fermions and bosons, interacting through pure hard cores, by Yang and Huang (H6, Y3).

Consider first the S-state interaction of two particles with a hard core of radius r_c. The Schrödinger equation is ($m = \hbar = 1$, r relative coordinate).

$$(\nabla^2 + k^2)\psi = 0 \qquad r > r_c \tag{4.1}$$

$$\psi(\mathbf{r}) = 0 \qquad r \leqslant r_c \ . \tag{4.2}$$

In virtue of (2) the wave function is *known* in region $r < r_c$. In the rest of the space the wave equation (1) holds. The chief difficulty in solving (1) is that it is not valid throughout the space. This difficulty may be removed if we replace (1) by an equation which is valid throughout the space and which has a solution ψ_{ps} (pseudo-wave function) which is exactly equal to ψ of (1) for $r > r_c$. To find this extended equation satisfied by ψ_{ps} we assume that

$$(\nabla^2 + k^2)\psi_{ps} = 0 \text{ (everywhere except at } \boldsymbol{r} = 0). \tag{4.3}$$

Since the wave function must vanish at $r = r_c$, the appropriate boundary condition would be

$$\psi_{ps}(\boldsymbol{r}) \underset{r\to 0}{\to} C(1 - r_c/r) \tag{4.4}$$

which implies that

$$r^2 \frac{\partial \psi_{ps}}{\partial r} \underset{r\to 0}{\to} r_c C \tag{4.5}$$

where the constant C is given by

$$C = \frac{\partial}{\partial r}[r\psi_{ps}(\boldsymbol{r})]_{r=0}. \tag{4.6}$$

On integrating both sides of (4.5) over the whole solid angle and converting to volume integrals over a small sphere one has

$$\int \frac{\partial^2}{\partial r^2}\psi_{ps}\,\mathrm{d}\boldsymbol{r} \to \int 4\pi r_c C\delta(\boldsymbol{r})\,\mathrm{d}\boldsymbol{r}.$$

Thus an alternative form of the boundary condition is obtained on equating the integrands of the above equation

$$\nabla^2\psi_{ps}(\boldsymbol{r}) \to 4\pi r_c \delta(\boldsymbol{r}) \frac{\partial}{\partial r}(r\psi_{ps}).$$

Whence the extended wave-equation may be obtained on comparing with equation (3).

$$(\nabla^2 + k^2)\psi_{ps}(\boldsymbol{r}) = 4\pi r_c \delta(\boldsymbol{r}) \frac{\partial}{\partial r}(r\psi_{ps}). \tag{4.7}$$

The term on the right hand side is the so called pseudo-potential for the S-state

$$V_{ps} = 4\pi r_c \delta(\boldsymbol{r}) \frac{\partial}{\partial r} r. \tag{4.8}$$

The exact pseudo-potential for arbitrary angular momentum and energy may be derived using phase-shifts and partial wave expansion of $\psi(\boldsymbol{r})$. The extended wave equation in the general case (H6, Y3) is given by

$$(\nabla^2 + k^2)\psi_{ps}(\boldsymbol{r}) = \left[-\frac{4\pi}{k \cot \eta_0}\right] \delta(\boldsymbol{r}) \frac{\partial}{\partial r}(r\psi_{ps})$$
$$+ \sum_{l=1}^{\infty} \sum_{m=-l}^{l} \left\{-\frac{\tan \eta_l}{k^{2l+1}} \frac{(2l-1)!!(l+1)}{l!\,2^l}\right\} Y_{lm}$$
$$\times \left[\frac{\delta(\boldsymbol{r})}{r^{l+2}} \left(\frac{\mathrm{d}}{\mathrm{d}r}\right)^{2l+1} (r^{l+1}(\psi_{ps})_{lm})\right] \tag{4.9}$$

where η_l are the phase-shifts for the scattering of the lth partial wave from the hard core and the other symbols have the usual meanings. The wave function ψ_{ps} can now be determined by a perturbation treatment of (9). The solution of (1) follows because $\psi_{ps} = \psi$ for $(r > r_c)$. By expanding appropriate terms it can be seen that the coefficient of ψ_{lm} on the right hand side of (9) starts with r_c^{2l+1}, so that for an accuracy up to second order in r_c it is sufficient to consider the simple pseudo-potential of Eq. (8). Since

$$-\frac{1}{k \cot \eta_0} = \frac{\tan ka}{k} = a[1 + \tfrac{1}{3}(ka)^2 + \ldots]$$

it is seen that the pseudo-potentials are valid only for $(ka) \ll 1$. Here a is the scattering length which in the present case is equal to the core radius [54] r_c.

In extending the method to the many-body problem we have to consider the equations $(2m_i = 1)$

$$\left(\sum_{i=1}^{N} \nabla_i^2 + \mathscr{E}\right)\Psi(\boldsymbol{r}_1 \ldots \boldsymbol{r}_n) = 0; \text{ for } \textit{all } |\boldsymbol{r}_i - \boldsymbol{r}_j| > r_c \tag{4.10}$$

$$\Psi(\boldsymbol{r}_1 \ldots \boldsymbol{r}_n) = 0; \text{ for } \textit{any } |\boldsymbol{r}_i - \boldsymbol{r}_j| \leqslant r_c. \tag{4.11}$$

[54] By general usage r_c is the core radius. Yang and Huang (H6, Y3) call it the diameter of the hard spheres which is perhaps more appropriate.

In the first approximation one would put the many body pseudo-potential equal to a sum of two-body pseudo-potentials of (7) or (9). This approximation takes into account the boundary condition (11) for only those regions of $3N$ dimensional space in which at most only two particles come close ($r_{ij} \leqslant r_c$) to each other.

If we want to satisfy the boundary condition for those regions in which three particles come close together, we must clearly have a three-body pseudo-potential V_3 to be obtained by solving the three-body problem given by

$$(\nabla_1^2 + \nabla_2^2 + \nabla_3^2 + \mathscr{E})\Psi = \delta(\boldsymbol{r}_1 - \boldsymbol{r}_2)\delta(\boldsymbol{r}_2 - \boldsymbol{r}_3) V_3 \Psi + \text{two body-terms.}$$

It is seen that the operator V_3 must have the dimensions of (length)^4. Since for $\mathscr{E} \to 0$ the only length available is r_c we conclude that $V_3 \sim r_c^4$. In general the n-body pseudo-potential will be proportional to r_c^{3n-5}. This argument, of course, does not tell us the relative magnitudes of higher order terms. For that purpose one has to calculate higher order terms and it is clear that the difficulty of obtaining higher order pseudo-potentials and their effect on the energy, increases in a highly non-linear fashion.

The expansion parameter in the two-body problem was (ka); in the many body problem there is the additional parameter of density ρ (or Fermi momentum k_F for fermion systems). It might be anticipated that the effective expansion parameter in the many body case will be $r_c k_F$ or $(r_c^3 \rho)^{\frac{1}{3}}$. Then it becomes evident that what we have here is basically a low density theory. In particular for an accuracy up to $(r_c k_F)^2$ one has to consider only the S-state two-body pseudo-potentials. The effective Hamiltonian in this case is

$$H = H_0 + \sum_{i<j} (v_{ps})_{ij}$$

$$= -\sum_{i=1}^{N} \nabla_i^2 + 16\pi r_c \sum_{i<j} \delta(\boldsymbol{r}_i - \boldsymbol{r}_j) \frac{\partial}{\partial r_{ij}} r_{ij}. \tag{4.12}$$

This may be solved by a perturbation method based on H_0. The result up to second order is (D7, Y3)

$$\frac{\mathscr{E}}{N} = \frac{k_F^2}{2M} \left[\frac{3}{5} + \frac{2}{\pi} (k_F r_c) + \frac{12}{35\pi^2} (11 - 2 \ln 2)(k_F r_c)^2 + \ldots \right]. \tag{4.13}$$

The same result may be derived by the t-matrix method (L4).

The method is more easily applicable to boson systems where no difficulties due to the Pauli principle arise. In the low density limit for fermions it is equivalent to the first Brueckner approximation (Ch. IV, Sec. 1). It does not seem to offer any obvious economy of labour in calculation of higher order effects for fermions. However, as one of the methods for determining three particle correlations, it might still be worthy of attention.

5. THEORY OF SUPERCONDUCTIVITY

The t-matrix and propagator modification that we have discussed correspond to partial summations of certain classes of terms in the perturbation series. First order t-matrix theory represents the effects of exactly taking into account the interaction between pairs of particles. In particular it neglects correlations of more than two particles and the correlations between the pairs (hence the name independent pair approximation G4). The former may be investigated by calculating three particle clusters in t-matrix expansion or by using one of the methods discussed in the preceding sections. The discussion of the latter in terms of perturbation theory is not easy. The importance of such correlations has been demonstrated by using variational and canonical transformation methods (B1, B2, B14, S14). Indeed such correlations are responsible for the of phenomenon of superconductivity in metals.

We shall here give a brief summary of various approaches to the problem of establishing a connection between perturbation theory and the theory of superconductivity. Basic to the latter is the fact that electron interactions with the lattice give rise to an *effective attraction between the electrons*. Hence the lattice plays no explicit role in the discussion. The problem of a nucleus is therefore very similar except in as much as the nucleon-nucleon interaction is not purely attractive and the nuclei have finite size. In the nuclear matter approximation the analogy is very close indeed.

The method of Bardeen, Cooper and Schrieffer (B14, S14) is perhaps more suited for studying the connection to the perturbation theory. It is a variational method in which the trial wave function is expressed in second notation of quantisation as

$$\Psi_0 = \prod_{\text{all}\,k} (\sqrt{1-h_k} + \sqrt{h_k}\,\eta^*_{k\uparrow}\eta^*_{-k\downarrow})|0_A)$$
$$= \prod_{k>k_F} (\sqrt{1-h_k} + \sqrt{h_k}\,\eta^*_{k\uparrow}\eta^*_{-k\downarrow})$$
$$\times \prod_{k'<k_F} (\sqrt{1-h_{k'}}\,\eta^*_{k'\uparrow}\eta^*_{-k'\downarrow} + \sqrt{h_{k'}})|\Phi_0) \qquad (5.1)$$

where h_k is a function of k to be determined by minimising energy, $|0_A)$ is the absolute vacuum with no particles, $|\Phi_0)$ is, as usual, the ground state of non-interacting particles – the fermi sea, corresponding to a given number N of particles and $\eta^*_{k\uparrow}$ $(\eta^*_{k\downarrow})$ are creation operators for particles with spin up (down) and momentum k.

This wave function has components corresponding to states in which pairs of particles exist with zero total momentum and opposite spins. Moreover, the total number of particles is not fixed [55] so that a constraint of fixed average total number of particles has to be included in the variational principle for determining h_k. To gain an insight into the structure of this wave function one should study the coordinate representation of its projection onto the space of N particles [56].

$$\Psi_0 = \sum_\alpha \mathscr{A}\{\varphi(1,2)\varphi(3,4)\ldots\times\ldots\bar{\varphi}(\bar{1},\bar{2})\bar{\varphi}(\bar{3},\bar{4})\ldots\} \qquad (5.2)$$

where there are 2α particles excited in pair functions φ, and 2α holes excited in pair functions $\bar{\varphi}$. $\mathscr{A}$ is the antisymmetriser. φ and $\bar{\varphi}$ are given by

$$\varphi(1,2) = \sum_{k>k_F} \sqrt{\frac{h_k}{1-h_k}}\,e^{ik\cdot x_{12}}\,\delta(1,\uparrow)\delta(2,\downarrow) \qquad (5.3)$$

$$\bar{\varphi}(\bar{1},\bar{2}) = \sum_{k<k_F} \sqrt{\frac{h_k}{1-h_k}}\,e^{ik\cdot x_{\bar{1},\bar{2}}}\,\delta(\bar{1},\uparrow)\delta(\bar{2},\downarrow) \qquad (5.4)$$

$\delta(s,\uparrow)$ and $\delta(s,\downarrow)$ are spin wave functions.

The wave function (1) was constructed to maximise the influence

[55] It has been shown that the effects of this fluctuation in the number of particles on ensemble averages of the quantities of interest are small for a system with a large number of particles. The question of principle behind this violation of the law of 'conservation of particle number' is discussed by Lipkin (L5).

[56] The distribution in number space is sharply peaked about N under the constraint that the average number of particles is N.

of interaction matrix elements of the type $(-k'\downarrow, k'\uparrow|H_{\text{int}}|-k\downarrow, k\uparrow)$. Suppose for a moment that

$$(-k'\downarrow, k'\uparrow|H_{\text{int}}|-k\downarrow, k\uparrow) = -V; \text{ for } k_{\text{F}} - \omega \leqslant k, k' \leqslant k_{\text{F}} + \omega$$
$$= 0 \quad \text{otherwise.} \tag{5.5}$$

and that there are N discrete states between $k_{\text{F}} + \omega$ and $k_{\text{F}} - \omega$ (ω small). We have, in other words, an N-dimensional matrix with all elements equal to $-V$. This has one eigenvalue $(-VN)$ and all other eigenvalues are zero. Thus, among the degenerate pair states, we have one state which has been depressed anomalously while the rest are unaffected (K9).

The actual situation as it emerges from superconductivity theory is, of course, more complicated but it preserves some similarity to the simple example above. It is found that if the interaction (5) is predominantly attractive, then the wave function (1) gives a state which is (anomalously) depressed compared to other (so-called 'normal') states. The rest of the interaction Hamiltonian in effect makes equal contributions to both 'normal' and superconducting states. It is a result of the minimisation procedure and also of canonical transformation methods, that the wave function and energy difference between superconducting and normal states depend non-analytically on the interaction. Even in simpler cases non-analytic perturbations are not fully understood (see, however, R2), but in view of the formal completeness of the wave function system of the zero-order Hamiltonian one expects that there would be a class of terms in perturbation theory whose sum – by analytic continuation or some other suitable device – will give rise to the anomalous behaviour. We shall return to this point later.

The wave function (1) represents a state in which particles are mostly in pairs which are 'bound', i.e., the pair wave functions (3) and (4) are localised. For electrons in superconducting metal, these wave functions decay in a distance of 10^{-4} cm, a distance in which there are present about 10^6 other pairs. Thus, although the particles are bound in pairs, the pairs themselves are not independent. They are continually 'rescattering' into each other because of the interaction (5). Hence it would seem that the chief difference between this 'superconducting' state and the ground state of Brueckner theory, resides in the fact that in the latter the pairs are unbound (Ch. IV,

Sec. 3d) and independent of each other. This is a rather important difference. The superconducting state in the limit of large numbers of particles is actually orthogonal to the unperturbed ground state (Fermi sphere), while the perturbed ground state accessible to ordinary perturbation theory (the Brueckner type ground state) is not. There exists no pair correlation in the latter and for that reason it is known as the normal state [57]. From the theory of superconductivity we know that the normal state lies above the superconducting state for temperatures T below the critical temperature T_c. In particular, it has been shown (B14, C5) that bound pairs are formed, the superconducting state exists and is lower than the normal state at $T = 0$ for all purely attractive forces between fermions irrespective of their strength.

These results are not directly applicable to the nuclear case because the nuclear interactions are singular, partly repulsive and include exchange forces. It immediately suggests itself that H_{int} in matrix elements (5) should be an effective interaction operator rather than the actual interaction v. In fact we are inclined to use the t-matrix as the interaction which is responsible for these pair correlations. This may be justified by recalling the discussion of Ch. IV, Secs. 2a and 7b. In Ch. IV, Sec. 2a it was pointed out that a straightforward evaluation of the t-matrix would lead to singularities. These singularities arise because of the vanishing of the energy denominators for some intermediate states; they are similar to the singularities that may arise from an incorrect handling of the states degenerate with the chosen configuration. In Ch. IV, Sec. 7, it was shown that a consistent way of treating such cases was to leave out the offending intermediate states from t-matrix calculations, use the *finite* t-matrix so obtained to replace the singular interaction v in the subspace spanned by the states in question and diagonalise the sub-matrix using some appropriate method. In the present case this justifies the use of t-matrix for H_{int} in (5) and the use of the methods of superconductivity theory for diagonalisation. All works on superfluidity [58]

[57] Actually a more precise definition of the normal state is needed. It has been provided recently by Klein (K16) in terms of solutions of an implicit equation involving the level shift operator.

[58] Superconductivity in metals arises from the superfluidity of electrons. To some it seems more appropriate to call the pair-correlated nuclear state 'super-

of nuclear matter and nuclei are based on this general idea (B1, B2, B32, B33, C6, M11, S9) although it is not quite explicitly stated (see, however, B32, B33). In spite of many unevaluated approximations and arbitrary choices for effective interaction used in these works, it seems to be generally agreed, as may be expected from the purely attractive nature of effective t-interaction, that at $T = 0$ a superfluid state exists and is lower than the normal ground state.

Brueckner and his coworkers (B33, B35) believe that the contribution of such effects is very small to the ground state total energy of nuclear systems.

This seems to patch up the theory, in as much as no glaring numerical inconsistencies remain and the values obtained may actually be quite close to the observed ones, but the real question of the connection between superconductivity theory and perturbation theory remains.

In a sense there is no conceptual difficulty. If we think of Brillouin-Wigner perturbation theory, it is quite easy to *imagine* the possibility that under certain circumstances the lowest root of the polynomial equation will depend non-analytically on the coupling constant. However, it will not be easy to investigate the behaviour of the root as the order of the polynomial is increased. In the case of the Rayleigh-Schrödinger series, one can imagine that if the series could be summed up in a closed form, then, depending on the strength of the interaction, one could have an analytic or non-analytic dependence on the coupling constant. Katz (K5) has constructed models of non-interacting fermion systems in which the above expectation is explicitly fulfilled. He has shown that in the analogue of the t-matrix or ladder approximation the energy of a many Fermion system which interacts only with an external potential has poles at those values of interaction strength where the exact energy has a finite discontinuity. This singularity is accompanied by the fact that on one side the bound solutions (Cooper type solution (C5)) of the analogue of the Bethe-Goldstone equation give lower energy, while on the other side the unbound solutions (Brueckner type) lead to lower energy. It is in-

fluid' because there is no question of 'conduction' in nuclei. However, there is really nothing to choose, except perhaps in the case of nuclear matter, because there is no question of a 'flow' either.

terpreted to mean that the true energy expression is a double valued function and the t-matrix approximation gives us only one branch of this function. In this connection Katz (K5) emphasises the contribution of the graphs which violate the Pauli principle but does not give an explicit enumeration of the important ones. Apart from the fact that these conclusions are based on models whose connection with the actual problem is rather vague, this work does not seem to touch upon the question that, as it seems, ordinary t-matrix would give an infinite result if the procedure (B33) sketched above is not adopted, irrespective of the strength of interaction. It would seem more profitable to discuss the meaning and correctness of the wave function and energies obtained by consistent use of the mixed procedure (B33) discussed above.

The procedure may also be inverted (D6). That is, one may first apply the canonical transformation of the theory of superconductivity and then use the t-matrix procedure on the transformed system.

Another approach to the present problem would be to use the perturbation theory of systems at non-zero temperatures [59] (B7, T9) and study the behaviour of various terms as the temperature goes to zero. In such a study Thouless (T9) finds, among other things, that the condition for temperature dependent ladder diagrams to give a convergent sum is identical with the condition for the temperature to be above the critical temperature.

Bloch and De Dominicis (B7) observe that in the limit of zero temperature there appear non-analytic contributions which seem to come from Cooper type solutions of the Bethe-Goldstone equation.

It is clear, however, that the connection of perturbation theory to the superconductivity theory has not been fully worked out.

[59] Properties of systems at non-zero temperatures may be calculated from the grand partition function Z given by

$$Z = \mathrm{Tr}\,(u(\beta));\ u(\beta) = \exp\,(-\beta H)$$

where $\beta = (kT)^{-1}$. The differential equation satisfied by u is

$$\partial u/\partial\beta = -Hu.$$

If β were (it) this would be the Schrödinger equation. Hence most methods of perturbation theory and diagrammatic methods may be taken over to develop expansions of the quantity Z. However, there exist minor differences in the expansions obtained by different authors (B7, B37, K1, K3, L1, M12, M13, T9, W14 and numerous other works quoted in these papers, particularly in M12).

LIST OF IMPORTANT SYMBOLS IN CHAPTER V

(N.B. *Other less frequently used symbols have been explained where they occur. Rest as in earlier chapters*)

$C_{\mu 0}$	amplitude of μth unperturbed state in the perturbed state Ψ_0. Matrix element of the wave-matrix M, (2.2, 2.3)
D	diagonal operator defined by the elements of the perturbed Hamiltonian w.r.t. to the unperturbed state, (2.5)
$\mathscr{D}$	determinant formed by the single particle wave functions of the chosen configuration, (3.1)
$f(r_{ij})$	state independent correlation function used in the trial wave function, (1.1, 1.2)
$f_{l_1 l_2}(12)$	correlation function for particle 1 in state l_1, 2 in state l_2, (1.4)
$G_{\mu\nu}$	operator connected to level shift, (2.6)
$N_{\mu\nu}$	operator defined by non-diagonal elements of the perturbed Hamiltonian, (2.5)
P_l	permutation operator acting on indices l, (1.3)
Π	product of the main diagonal elements of $\mathscr{D}$
$\Pi^i \Pi^{ij} \ldots$	products obtained by omitting $\varphi^i(i)$, $\varphi^i(i)\varphi^j(j)$, ... from Π
φ^i, φ^{mi}	unperturbed single particle wave functions
φ_i, φ_{m_i}	modified single particle wave function, (3.3a)
$\varphi^{k_i k_j}$	antisymmetric unperturbed two-particle wave function (3.7c)
$\Phi^{k_1 k_2 k_3 \ldots}_{m_1 m_2 m_3 \ldots}$	an excited configuration obtained from the chosen configuration by emptying states m_1, m_2, m_3, ... inside the chosen configuration and filling the states k_1, k_2, k_3, ... outside it
ψ_{ij}, ψ_{ijk}, $\psi_{ijk \ldots}$	correlated wave functions for successively large numbers of particles in the system

REFERENCES

(N.B. *Italicized numbers in square brackets refer to pages of the present book*)

A1. J. Ashkin and T. Y. Wu, Phys. Rev. **73** (1948) 973 [*153, 155*]
A2. J. B. Aviles, Ann. of Phys. **5** (1958) 251 [*201*]
A3. Alder *et al.*, Rev. Mod. Phys. **28** (1956) 432 [*197*]
A4. N. Austern and P. Iano, Nucl. Phys. **18** (1960) 672 [*203*]

B1. Bogoliubov, Tolmachev and Shirkov, A New Method in the Theory of Superconductivity (Consultants Bureau, New York, 1959); also author's mimeographed version and articles in Nuovo Cimento and Zhur. Theor. Exptl. Phys. (Russian) [*1, 113, 118, 215, 219*]
B2. S. T. Belyaev, Mat. Fys. Medd. Dan. Vid. Selsk. **31**, No. 11 (1959); *also in* ref. D3, p. 343 [*1, 215, 219*]
B3. D. Bohm, in ref. D3, p. 401 [*1*]
B4. D. Bohm and E. P. Gross, Phys. Rev. **75** (1949) 1851, 1864 [*1*]
B5. D. Bohm and D. Pines, Phys. Rev. **85** (1951) 338; **92** (1953) 609 [*1*]
B6. A. Bohr and B. R. Mottelson, Mat. Fys. Medd. Dan. Vid. Selsk. **30**, No. 1 (1956) [*1*]
B7. C. Bloch and C. De Dominicis, Nucl. Phys. **7** (1958) 459; **10** (1959) 181, 509; C. Bloch, in ref. D3, p. 257 [*2, 220*]
B8. W. Brenig, Nucl. Phys. **4** (1957) 363 [*180, 207*]
B9. L. Brillouin, J. Phys. Radium **4** (1933) 1 [*11*]
B10. M. Bolsterli and E. Feenberg, Phys. Rev. **101** (1956) 1349 [*14, 181, 190*]
B11. K. A. Brueckner, Phys. Rev. **100** (1955) 36 [*35, 36, 49, 118*]
B12. K. A. Brueckner and C. Levinson, Phys. Rev. **97** (1955) 1344 [*35, 118, 195*]
B13. K. A. Brueckner, in ref. D3, p. 42 [*49, 53, 144, 147, 159, 160, 161, 192*]
B14. Bardeen, Cooper and Schrieffer, Phys. Rev. **108** (1957) 1593 [*113, 215, 218*]
B15. H. A. Bethe, Phys. Rev. **103** (1956) 1353 [*45, 135, 138, 151, 153, 194, 195, 211*]
B16. H. A. Bethe and J. Goldstone, Proc. Roy. Soc. A **238** (1957) 551 [*149, 160, 161, 162*]
B17. Brueckner, Gammel and Weitzner, Phys. Rev. **110** (1958) 431 (abbreviated as BGW in text) [*154, 166, 170, 171, 174, 175, 180, 181, 184, 194*]
B18. K. A. Brueckner, in Proc. International Conf. Nucl. Optical Model, Florida State University, Talahassee (1959), and in Proc. Int. Conf. on

Nucl. Structure, Kingston, (Toronto University Press, and North-Holland Publishing Co., Amsterdam, 1960) [*180, 194*]

B19. K. A. Brueckner and J. L. Gammel, Phys. Rev. **109** (1958) 1023 [*131, 134, 152, 159, 160, 161, 165, 166, 177, 181, 182, 186*]

B20. J. M. Blatt and C. J. Biedenharn, Phys. Rev. **86** (1952) 399 [*153*]

B21. J. M. Blatt and V. Weisskopf, Theoretical Nuclear Physics (Wiley, New York, 1952) [*153, 211*]

B22. Brueckner, Levinson and Mahmood, Phys. Rev. **95** (1954) 217 [*146, 159*]

B23. K. A. Brueckner and J. L. Gammel, Phys. Rev. **109** (1958) 1041 [*142, 165*]

B24. R. Brout, Phys. Rev. Letters **5** (1960) 193 [*172*]

B25. K. A. Brueckner and D. Goldman, Phys. Rev. **116** (1959) 424 [*180, 182*]

B26. K. A. Brueckner, Phys. Rev. **110** (1958) 597 [*176, 177, 182*]

B27. K. A. Brueckner, Phys. Rev. **97** (1955) 1353 [*146, 193*]

B28. Brueckner, Eden and Francis, Phys. Rev. **98** (1955) 1445 [*143, 194*]

B29. Bell, Eden and Skyrme, Nucl. Phys. **2** (1956) 586; J. S. Bell, Nucl. Phys. **4** (1957) 295 [*195*]

B30. A. Bohr, Mat. Fys. Medd. Dan. Vid. Selsk. **26**, No. 14 (1952); A. Bohr and B. R. Mottelson, *ibid.*, **27**, No. 16 (1953) [*197*]

B31. K. A. Brueckner and R. Thieberger, Phys. Rev. Letters **4** (1960) 466 [*197*]

B32. K. A. Brueckner and K. Sawada, Phys. Rev. **106** (1957) 1117, 1128 [*135, 142, 219*]

B33. Brueckner, Soda, Anderson and Morel, Phys. Rev. **118** (1960) 1442 [*134, 153, 219, 220*]

B34. Brueckner, Gammel and Kubis, Phys. Rev. **118** (1960) 1438 [*177, 179*]

B35. Brueckner, Lockett and Rotenberg, Phys. Rev. **121** (1961) 255 [*136, 153, 175, 180, 194, 219*]

B36. K. A. Brueckner and D. Goldman, Phys. Rev. **117** (1960) 207 [*179*]

B37. V. L. Bonch-Bruevich and Sh. M. Kogan, Ann. of Phys. **9** (1960) 125 [*2, 220*]

B38. J. Bardeen, Phys. Rev. **51** (1937) 799 [*150*]

B39. R. J. Blin-Stoyle, Phil. Mag. **46** (1955) 973 [*174*]

B40. F. C. Barker, Phys. Rev. **122** (1961) 572 [*196*]

C1. E. U. Condon and G. S. Shortly, Theory of Atomic Spectra (Cambridge University Press, 1953) [*6, 191, 195*]

C2. J. S. R. Chisholm and E. J. Squires, Nucl. Phys. **13** (1959) 156 [*123, 152, 207*]

C3. J. W. Clark and E. Feenberg, Phys. Rev. **113** (1959) 388 [*201*]

C4. F. Coester and H. Kümmel, Nucl. Phys. **17** (1960) 477 [*207*]

C5. L. N. Cooper, Phys. Rev. **104** (1956) 1189; and Am. J. Phys. **28** (1960) 91 [*218, 219*]

C6. Cooper, Mills and Sessler, Phys. Rev. **114** (1959) 1377 [*153, 219*]

C7. J. W. Clark, Ann. of Phys. **11** (1960) 483 [*175, 181*]

C8. L. N. Cooper, Phys. Rev. **122** (1961) 1021 [*73*]

D1. A. Dalgarno and J. T. Lewis, Proc. Roy. Soc. A **233** (1956) 70 [*27, 29, 30*]
D2. A. Dalgarno and L. Stewart, Proc. Roy. Soc. A **238** (1956) 269 [*27, 29, 30*]
D3. B. De Witt, editor, Many Body Problem (Methuen, 1959) (see refs. B2, B3, B7, B13, H2, H6, S14 and T6)
D4. B. De Witt, Phys. Rev. **103** (1956) 1565 [*146*]
D5. J. Dabrowski, Proc. Phys. Soc. A **71** (1958) 659 [*201*]
D6. B. B. Dotsenko, in Proc. Int. Conf. on Nucl. Structure, Kingston (Toronto University Press, and North-Holland Publishing Co., Amsterdam, 1960) [*220*]
D7. C. De Dominicis and P. C. Martin, Phys. Rev. **105** (1957) 1419 [*214*]
D8. W. E. Drummond, Phys. Rev. **116** (1960) 183 [*182*]

E1. R. J. Eden and N. C. Francis, Phys. Rev. **97** (1955) 1366 [*11, 12, 195*]
E2. E. T. Epstein, Am. J. Phys. **28** (1960) 495 [*14*]
E3. R. J. Eden, Proc. Roy. Soc. A **235** (1956) 408 [*139*]
E4. R. J. Eden and V. J. Emery, Proc Roy. Soc. A **248** (1948) 266; Eden, Emery and Sampanthar, *ibid.*, A **253** (1959) 177, 186 [*138, 139, 148, 163, 180, 186, 187, 191, 194*]
E5. V. J. Emery, Nucl. Phys. **12** (1959) 69; **19** (1960) 154 [*123, 152*]
E6. J. P. Elliott and T. H. R. Skyrme, Proc. Roy. Soc. A **232** (1955) 561 [*181, 191*]

F1. E. S. Fradkin, Nucl. Phys. **12** (1959) 465 [*2*]
F2. N. Fukuda, Phys. Rev. **103** (1956) 420 [*12*]
F3. E. Feenberg, Phys. Rev. **103** (1956) 1116 [*14, 16, 17, 19, 21*]
F4. E. Feenberg and P. Goldhammer, Phys. Rev. **105** (1957) 750 [*14, 15, 17*]
F5. E. Feenberg, Ann. of Phys. **3** (1958) 292 [*14, 21, 23, 25*]
F6. E. Feenberg, Phys. Rev. **74** (1948) 206 [*32, 33*]
F7. N. Fukuda and R. Newton, Phys. Rev. **103** (1956) 1558 [*146*]
F8. E. Feenberg, Phys. Rev. **59** (1941) 149, 593; Rev. Mod. Phys. **19** (1947) 239 [*192*]
F9. E. Fermi, Nuclear Physics (Chicago University Press, Chicago, 1950) p. 23 [*192*]
F10. W. E. Frahn and R. H. Lemmer, Nuovo Cimento **5** (1957) 1564 and **6** (1957) 664 [*194*]
F11. E. Fermi, Ricera Sci. **7** (1936) 13 [*211*]
F12. J. Fujita, Nucl. Phys. **14** (1960) 648 [*201*]
F13. N. C. Francis and K. M. Watson, Phys. Rev. **92** (1955) 291 [*31, 34, 35*]
F14. L. L. Foldy and W. Tobocman, Phys. Rev. **105** (1957) 1099 [*197*]
F15. A. M. Feingold, Phys. Rev. **101** (1956) 258; **105** (1957) 944 [*174*]

G1. J. Goldstone, Proc. Roy. Soc. A **239** (1957) 267 [*45, 66, 70*]
G2. P. Goldhammer and E. Feenberg, Phys. Rev. **101** (1956) 1233 [*14, 17, 19*]

G3. M. Gell-Mann and F. Low, Phys. Rev. **84** (1951) 350 [*45, 66*]
G4. Gomes, Walecka and Weisskopf, Ann. of Phys. **3** (1958) 241 [*162, 164, 215*]
G5. J. L. Gammel and R. M. Thaler, Phys. Rev. **107** (1957) 1337 [*165*]
G6. E. P. Gross, Phys. Rev. Letters **4** (1960) 599 [*171*]
G7. Glassgold, Heckrotte and Watson, Ann. of Phys. **6** (1959) 1 [*172, 197*]
G8. A. E. S. Green, Rev. Mod. Phys. **30** (1958) 569; A. E. S. Green and P. C. Sood, Phys. Rev. **111** (1958) 1147 [*194*]
G9. L. C. Gomes, Phys. Rev. **116** (1959) 1226 [*198*]
G10. M. Gell-Mann and K. A. Brueckner, Phys. Rev. **106** (1957) 364 [*118, 135*]

H1. N. M. Hugenholtz, Physica **23** (1957) 481 [*45, 73, 76, 83, 95*]
H2. N. M. Hugenholtz, in ref. D3, p. 1 [*45, 73, 76, 77, 83*]
H3. N. M. Hugenholtz, Physica **23** (1957) 533 [*76, 108*]
H4. N. M. Hugenholtz and L. van Hove, Physica **24** (1958) 363 [*45, 106, 112, 138, 172, 176, 177*]
H5. E. L. Hill, in Encyclopedia of Physics, **39** (Springer, Berlin, 1957) p. 202 [*192*]
H6. K. Huang, in ref. D3, p. 601 [*211, 213*]
H7. D. R. Hartree, The Calculation of Atomic Structure (Wiley, New York, 1957) [*191*]
H8. V. Heine, Group theory in quantum mechanics (Pergamon Press, 1960) p. 41 [*6*]
H9. J. Hughes and K. J. Le Couteur, Proc. Phys. Soc. A **63** (1950) 1219 [*174*]
H10. E. M. Henley and Th. W. Ruizgrok, Ann. of Phys. **12** (1961) 409; E. M. Henley and L. Wilets, Ann. of Phys. **14** (1961) 120 [*172*]

I1. F. Iwamoto and M. Yamada, Prog. Theor. Phys. **17** (1957) 543; **18** (1957) 345; **19** (1958) 599 [*44, 201*]

J1. M. W. Johnson and E. Teller, Phys. Rev. **98** (1955) 783 [*143*]
J2. R. Jastrow, Phys. Rev. **98** (1955) 1478 [*201*]
J3. B. Jancovici, Nucl. Phys. **21** (1960) 256 [*181, 198*]

K1. A. Klein and R. Prange, Phys. Rev. **112** (1958) 994, 1008 [*2, 112, 113, 176, 220*]
K2. T. Kato, Prog. Theor. Phys. **4** (1949) 514 [*76*]
K3. W. Kohn and J. M. Luttinger, Phys. Rev. **118** (1960) 41 [*72, 220*]
K4. A. Klein, Phys. Rev. Letters **4** (1960) 601 [*72*]
K5. A. Katz, Nucl. Phys. **20** (1960) 663 [*71, 219, 220*]
K6. R. A. Kromhout, Phys. Rev. **107** (1957) 215 [*203*]
K7. W. Kohn and S. J. Nettle, Phys. Rev. Letters **5** (1960) 8 [*172*]
K8. K. Kumar, Nucl. Phys. **21** (1960) 99 [*209, 211*]
K9. Katz, De Shalit and Talmi, Nuovo Cimento **16** (1960) 485 [*164, 217*]
K10. T. Kato, Trans. Am. Math. Soc. **70** (1951) 195 [*194*]

K11. L. A. Khalfin, Sov. Phys. JETP **6** (1958) 1053 (original Russian reference, Zhr. Theor. Exptl. Phys. **33** (1957) 1371; Doklady Nauk. S.S.R. **115** (1957) 277) [*106*]

K12. K. Kumar and M. A. Preston, Can. J. Phys. **33** (1955) 298 [*193*]

K13. K. Kumar and M. A. Preston, Phys. Rev. **107** (1957) 1099 [*193*]

K14. S. Kahana, Phys. Rev. **117** (1960) 123 [*142*]

K15. Kerman, McManus and Thaler, Ann. of Phys. **8** (1959) 551 [*146, 168, 197*]

K16. A. Klein, Phys. Rev. **121** (1961) 950, 957 [*112, 113, 176, 179, 218*]

K17. K. Kumar and R. K. Bhaduri, Phys. Rev. **122** (1961) 1926 [*183*]

K18. H. Kanazawa *et al.*, Prog. Theor. Phys. **23** (1960) 408, 426, 433 [*135*]

K19. J. Keilson, Phys. Rev. **82** (1951) 759 [*174*]

K20. H. S. Köhler, Ann. of Phys. **12** (1961) 444 [*135, 138*]

K21. K. L. Kowalski and D. Feldman, J. Math. Phys. **2** (1961) 499 [*153, 162*]

L1. J. M. Luttinger and J. C. Ward, Phys. Rev. **118** (1960) 1417 [*2, 72, 220*]

L2. H. J. Lipkin, Phys. Rev. **109** (1958) 2071 [*181, 190*]

L3. G. Lüders, Z. Naturf. **14a**, (1959) 1 [*164*]

L4. Levinger, Razavy, Rojo and Webre, Phys. Rev. **119** (1960) 230 [*150, 161, 163, 192*]

L5. H. J. Lipkin, Ann. of Phys. **9** (1960) 272 [*216*]

L6. G. Lüders, Z. Naturf. **16a** (1961) 76 [*191*]

L7. J. S. Lomont, Applications of finite groups (Academic Press, 1959) p. 177 [*6*]

L8. K. J. Le Couteur, Proc. Phys. Soc. A **63** (1950) 259 [*150*]

M1. P. M. Morse and H. Feshbach, Methods of Theoretical Physics (McGraw-Hill, 1953) Ch. 9 [*14, 32, 33*]

M2. H. J. Mang and W. Wild, Z. Phys. **154** (1959) 182 [*148, 180, 181, 191, 194*]

M3. M. L. Mehta, Nucl. Phys. **12** (1959) 333 [*123, 152*]

M4. A. N. Mitra and V. L. Narasimham, Nucl. Phys. **14** (1960) 407; A. N. Mitra and J. H. Naqvi, Nucl. Phys. **25** (1961) 307 [*151, 152, 167*]

M5. P. Mittelstaedt, Nucl. Phys. **17** (1960) 499 [*176, 179*]

M6. P. T. Matthews and A. Salam, Phys. Rev. **112** (1958) 283; **115** (1959) 1079 [*106*]

M7. S. A. Moszkowski and B. L. Scott, Ann. of Phys. **11** (1960) 65 [*150, 159, 163, 170, 177*]

M8. S. A. Moszkowski and B. L. Scott, Phys. Rev. Letters **1** (1958) 298 [*170*]

M9. M. L. Mehta, Nucl. Phys. **20** (1960) 533 [*124*]

M10. J. E. Mayer and M. G. Mayer, Statistical Mechanics (Wiley, 1940) [*44*]

M11. Mills, Sessler, Moszkowski and Shankland, Phys. Rev. Letters **3** (1959) 381 [*219*]

M12. P. C. Martin and J. Schwinger, Phys. Rev. **115** (1959) 1342 [*2, 220*]

M13. F. Mohling, Phys. Rev. **122** (1961) 1043, 1062 [*2, 220*]

N1. S. G. Nilsson, Mat. Fys. Medd. Dan. Vid. Selsk. **29**, No. 16 (1955) [*193*]
N2. R. K. Nesbet, Phys. Rev. **109** (1958) 1632 [*203, 205, 207*]
N3. B. P. Nigam and M. K. Sundaresan, Can. J. Phys. **36** (1958) 571 [*152, 174*]
N4. L. H. Nosanow, Physica **26** (1960) 1124 [*106*]

O1. A. W. Overhauser, Phys. Rev. Letters **4** (1960) 415, 462 [*171*]

P1. A. Pais and R. Serber, Phys. Rev. **105** (1957) 1636 [*1*]
P2. J. C. Polkinghorne, Nucl. Phys. **3** (1957) 94 [*44*]
P3. I. Ya. Pomeranchuk, Zhur. Theor. Exptl. Phys. **35** (1958) 524, (translation Sov. Phys. JETP. **35 (8)** (1959) 361) [*172*]
P4. S. P. Pandya and S. K. Shah, Nucl. Phys. **24** (1961) 326 [*196*]
P5. D. C. Peaslee, Phys. Rev. **124** (1961) 839 [*196*]

R1. W. Riesenfeld and K. M. Watson, Phys. Rev. **104** (1956) 492 [*3, 7, 31, 37, 134*]
R2. F. Rellich, Perturbation theory of Eigenvalue Problem, Mimeographed notes (Institute of Mathematics, New York University, 1953) [*7, 72, 217*]
R3. Ross, Mark and Lawson, Phys. Rev. **102** (1956) 1613 [*194*]
R4. Ross, Lawson and Mark, Phys. Rev. **104** (1956) 401 [*194*]
R5. L. S. Rodberg, Ann. of Phys. **2** (1957) 199; **9** (1960) 373 [*135*]

S1. K. Sawada, Phys. Rev. **119** (1960) 2090 [*66*]
S2. G. Speisman, Phys. Rev. **107** (1957) 1180 [*27*]
S3. C. Schwartz, Ann. of Phys. **6** (1959) 156, 170 [*27*]
S4. C. Schwartz and J. J. Tiemann, Ann. of Phys. **6** (1959) 178 [*27*]
S5. Schweber, Bethe and de Hoffmann, Mesons and Fields, Vol. **1** (Row, Peterson and Co., 1955) p. 196 [*45, 68*]
S6. M. H. Stone, Linear Transformations in Hilbert Space (Am. Math. Soc., New York, 1932) [*76*]
S7. J. Schwinger, Ann. of Phys. **9** (1960) 169 [*106*]
S8. G. L. Shaw, Ann. of Phys. **8** (1959) 509 [*108, 159, 197, 198*]
S9. V. G. Soloviev, Nucl. Phys. **9** (1959) 655 [*219*]
S10. J. Sawicki, Nucl. Phys. **13** (1959) 350 [*174, 175*]
S11. F. Seitz, The Modern Theory of Solids (McGraw-Hill, New York, 1940) Ch. VI [*171*]
S12. E. J. Squires, Nucl. Phys. **13** (1959) 224 [*172*]
S13. T. H. R. Skyrme, Phil. Mag. **1** (1956) 1043 [*183, 193*]
S14. J. R. Schrieffer, in ref. D8, p. 541 [*215*]
S15. A. J. F. Siegert, Phys. Rev. **116** (1959) 1057 [*66*]
S16. W. J. Swiatecki, Phys. Rev. **101** (1956) 1321; **103** (1956) 265 [*161, 192*]

T1. S. Tomonaga, Prog. Theor. Phys. **13** (1955) 467, 482 [*1*]
T2. H. Tanaka, Prog. Theor. Phys. **13** (1955) 497 [*12*]

T3. R. E. Trees, Phys. Rev. **102** (1956) 1553 [*14, 17*]
T4. A. Temkin, Ann. of Phys. **9** (1960) 93 [*44, 201, 202*]
T5. W. Tobocman, Phys. Rev. **107** (1957) 203[*119, 122, 138*]
T6. D. J. Thouless, in ref. D3, and other papers in Phys. Rev. [*123*]
T7. D. J. Thouless, Nucl. Phys. **21** (1960) 225 [*62, 172*]
T8. Tang, Lemmer, Wyatt and Green, Phys. Rev. **116** (1959) 402 [*194*]
T9. D. J. Thouless, Ann. of Phys. **10** (1960) 553 [*220*]
t.H1. D. ter Haar, Introduction to many body problem (Interscience Publ., New York, 1958) [*1*]

U1. T. Usui, Prog. Theor. Phys. **23** (1960) 787; T. Usui and E. Fujita, *ibid.*, **23** (1960) 799 [*118, 135*]

V1. C. Villi, Nuovo Cimento **10** (1958) 259; and Nucl. Phys. **9** (1959) 306 [*112, 176, 179*]
v.H1. L. Van Hove, Physica **21** (1955) 901; **22** (1956) 343 [*100, 105, 106*]
v.H2. L. Van Hove, M.I.T. report on anharmonic oscillations [*44*]
v.V1. J. H. Van Vleck, Phys. Rev. **48** (1935) 367 [*150*]

W1. E. P. Wigner, Math. u. Naturw. Anz. Ungar. Akad. Wiss. **53** (1935) 475 [*11*]
W2. K. M. Watson, Phys. Rev. **89** (1953) 575 [*31, 34, 35*]
W3. K. M. Watson, Phys. Rev. **105** (1957) 1388 [*31, 34*]
W4. E. P. Wigner, Phys. Rev. **94** (1954) 77 [*14*]
W5. K. Wildermuth and Th. Kanellopoulos, Nucl. Phys. **7** (1958) 150; **14** (1960) 349 [*44*]
W6. G. Wentzel, Phys. Rev. **120** (1960) 659 [*72*]
W7. V. Weisskopf and A. de Shalit, Ann. of Phys. **5** (1958) 282 [*202*]
W8. L. Wolfenstein and J. Ashkin, Phys. Rev. **85** (1952) 947; L. Wolfenstein, Ann. Revs. Nucl. Sciences **6** (1956) 43 [*168*]
W9. E. P. Wigner, Trans. Faraday Soc. **34** (1938) 678 [*171*]
W10. E. Werner, Nucl. Phys. **10** (1959) 688 [*159*]
W11. Wilets, Hill and Ford, Phys. Rev. **91** (1953) 1148 [*192*]
W12. A. H. Wapstra, *in* Encyclopedia of Physics, **38**/1 (Springer, Berlin, 1958) p. 1 [*192, 193*]
W13. L. Wilets, Rev. Mod. Phys. **30** (1958) 542 [*183, 193*]
W14. K. M. Watson, Phys. Rev. **103** (1956) 489 [*220*]
W15. P. A. Wolf, Phys. Rev. **116** (1959) 544 [*135*]

Y1. Young, Biedenharn and Feenberg, Phys. Rev. **106** (1957) 1151 [*14, 16, 17, 19, 21*]
Y2. Y. Yamaguchi and Y. Yamaguchi, Phys. Rev. **95** (1954) 1628 [*151, 152, 167*]
Y3. C. N. Yang and K. Huang, Phys. Rev. **105** (1957) 767 [*213, 214*]

AUTHOR INDEX

As throughout this book, indications like A3, B33, refer to pp. 222–228

SUBJECT INDEX

A CATALOG OF SELECTED

DOVER BOOKS

IN SCIENCE AND MATHEMATICS

Astronomy

CHARIOTS FOR APOLLO: The NASA History of Manned Lunar Spacecraft to 1969, Courtney G. Brooks, James M. Grimwood, and Loyd S. Swenson, Jr. This illustrated history by a trio of experts is the definitive reference on the Apollo spacecraft and lunar modules. It traces the vehicles' design, development, and operation in space. More than 100 photographs and illustrations. 576pp. 6 3/4 x 9 1/4. 0-486-46756-2

EXPLORING THE MOON THROUGH BINOCULARS AND SMALL TELESCOPES, Ernest H. Cherrington, Jr. Informative, profusely illustrated guide to locating and identifying craters, rills, seas, mountains, other lunar features. Newly revised and updated with special section of new photos. Over 100 photos and diagrams. 240pp. 8 1/4 x 11. 0-486-24491-1

WHERE NO MAN HAS GONE BEFORE: A History of NASA's Apollo Lunar Expeditions, William David Compton. Introduction by Paul Dickson. This official NASA history traces behind-the-scenes conflicts and cooperation between scientists and engineers. The first half concerns preparations for the Moon landings, and the second half documents the flights that followed Apollo 11. 1989 edition. 432pp. 7 x 10. 0-486-47888-2

APOLLO EXPEDITIONS TO THE MOON: The NASA History, Edited by Edgar M. Cortright. Official NASA publication marks the 40th anniversary of the first lunar landing and features essays by project participants recalling engineering and administrative challenges. Accessible, jargon-free accounts, highlighted by numerous illustrations. 336pp. 8 3/8 x 10 7/8. 0-486-47175-6

ON MARS: Exploration of the Red Planet, 1958-1978--The NASA History, Edward Clinton Ezell and Linda Neuman Ezell. NASA's official history chronicles the start of our explorations of our planetary neighbor. It recounts cooperation among government, industry, and academia, and it features dozens of photos from Viking cameras. 560pp. 6 3/4 x 9 1/4. 0-486-46757-0

ARISTARCHUS OF SAMOS: The Ancient Copernicus, Sir Thomas Heath. Heath's history of astronomy ranges from Homer and Hesiod to Aristarchus and includes quotes from numerous thinkers, compilers, and scholasticists from Thales and Anaximander through Pythagoras, Plato, Aristotle, and Heraclides. 34 figures. 448pp. 5 3/8 x 8 1/2. 0-486-43886-4

AN INTRODUCTION TO CELESTIAL MECHANICS, Forest Ray Moulton. Classic text still unsurpassed in presentation of fundamental principles. Covers rectilinear motion, central forces, problems of two and three bodies, much more. Includes over 200 problems, some with answers. 437pp. 5 3/8 x 8 1/2. 0-486-64687-4

BEYOND THE ATMOSPHERE: Early Years of Space Science, Homer E. Newell. This exciting survey is the work of a top NASA administrator who chronicles technological advances, the relationship of space science to general science, and the space program's social, political, and economic contexts. 528pp. 6 3/4 x 9 1/4. 0-486-47464-X

STAR LORE: Myths, Legends, and Facts, William Tyler Olcott. Captivating retellings of the origins and histories of ancient star groups include Pegasus, Ursa Major, Pleiades, signs of the zodiac, and other constellations. "Classic." – *Sky & Telescope.* 58 illustrations. 544pp. 5 3/8 x 8 1/2. 0-486-43581-4

A COMPLETE MANUAL OF AMATEUR ASTRONOMY: Tools and Techniques for Astronomical Observations, P. Clay Sherrod with Thomas L. Koed. Concise, highly readable book discusses the selection, set-up, and maintenance of a telescope; amateur studies of the sun; lunar topography and occultations; and more. 124 figures. 26 halftones. 37 tables. 335pp. 6 1/2 x 9 1/4. 0-486-42820-6

Chemistry

MOLECULAR COLLISION THEORY, M. S. Child. This high-level monograph offers an analytical treatment of classical scattering by a central force, quantum scattering by a central force, elastic scattering phase shifts, and semi-classical elastic scattering. 1974 edition. 310pp. 5 3/8 x 8 1/2. 0-486-69437-2

HANDBOOK OF COMPUTATIONAL QUANTUM CHEMISTRY, David B. Cook. This comprehensive text provides upper-level undergraduates and graduate students with an accessible introduction to the implementation of quantum ideas in molecular modeling, exploring practical applications alongside theoretical explanations. 1998 edition. 832pp. 5 3/8 x 8 1/2. 0-486-44307-8

RADIOACTIVE SUBSTANCES, Marie Curie. The celebrated scientist's thesis, which directly preceded her 1903 Nobel Prize, discusses establishing atomic character of radioactivity; extraction from pitchblende of polonium and radium; isolation of pure radium chloride; more. 96pp. 5 3/8 x 8 1/2. 0-486-42550-9

CHEMICAL MAGIC, Leonard A. Ford. Classic guide provides intriguing entertainment while elucidating sound scientific principles, with more than 100 unusual stunts: cold fire, dust explosions, a nylon rope trick, a disappearing beaker, much more. 128pp. 5 3/8 x 8 1/2. 0-486-67628-5

ALCHEMY, E. J. Holmyard. Classic study by noted authority covers 2,000 years of alchemical history: religious, mystical overtones; apparatus; signs, symbols, and secret terms; advent of scientific method, much more. Illustrated. 320pp. 5 3/8 x 8 1/2. 0-486-26298-7

CHEMICAL KINETICS AND REACTION DYNAMICS, Paul L. Houston. This text teaches the principles underlying modern chemical kinetics in a clear, direct fashion, using several examples to enhance basic understanding. Solutions to selected problems. 2001 edition. 352pp. 8 3/8 x 11. 0-486-45334-0

PROBLEMS AND SOLUTIONS IN QUANTUM CHEMISTRY AND PHYSICS, Charles S. Johnson and Lee G. Pedersen. Unusually varied problems, with detailed solutions, cover of quantum mechanics, wave mechanics, angular momentum, molecular spectroscopy, scattering theory, more. 280 problems, plus 139 supplementary exercises. 430pp. 6 1/2 x 9 1/4. 0-486-65236-X

ELEMENTS OF CHEMISTRY, Antoine Lavoisier. Monumental classic by the founder of modern chemistry features first explicit statement of law of conservation of matter in chemical change, and more. Facsimile reprint of original (1790) Kerr translation. 539pp. 5 3/8 x 8 1/2. 0-486-64624-6

MAGNETISM AND TRANSITION METAL COMPLEXES, F. E. Mabbs and D. J. Machin. A detailed view of the calculation methods involved in the magnetic properties of transition metal complexes, this volume offers sufficient background for original work in the field. 1973 edition. 240pp. 5 3/8 x 8 1/2. 0-486-46284-6

GENERAL CHEMISTRY, Linus Pauling. Revised third edition of classic first-year text by Nobel laureate. Atomic and molecular structure, quantum mechanics, statistical mechanics, thermodynamics correlated with descriptive chemistry. Problems. 992pp. 5 3/8 x 8 1/2. 0-486-65622-5

ELECTROLYTE SOLUTIONS: Second Revised Edition, R. A. Robinson and R. H. Stokes. Classic text deals primarily with measurement, interpretation of conductance, chemical potential, and diffusion in electrolyte solutions. Detailed theoretical interpretations, plus extensive tables of thermodynamic and transport properties. 1970 edition. 590pp. 5 3/8 x 8 1/2. 0-486-42225-9

Engineering

FUNDAMENTALS OF ASTRODYNAMICS, Roger R. Bate, Donald D. Mueller, and Jerry E. White. Teaching text developed by U.S. Air Force Academy develops the basic two-body and n-body equations of motion; orbit determination; classical orbital elements, coordinate transformations; differential correction; more. 1971 edition. 455pp. 5 3/8 x 8 1/2. 0-486-60061-0

INTRODUCTION TO CONTINUUM MECHANICS FOR ENGINEERS: Revised Edition, Ray M. Bowen. This self-contained text introduces classical continuum models within a modern framework. Its numerous exercises illustrate the governing principles, linearizations, and other approximations that constitute classical continuum models. 2007 edition. 320pp. 6 1/8 x 9 1/4. 0-486-47460-7

ENGINEERING MECHANICS FOR STRUCTURES, Louis L. Bucciarelli. This text explores the mechanics of solids and statics as well as the strength of materials and elasticity theory. Its many design exercises encourage creative initiative and systems thinking. 2009 edition. 320pp. 6 1/8 x 9 1/4. 0-486-46855-0

FEEDBACK CONTROL THEORY, John C. Doyle, Bruce A. Francis and Allen R. Tannenbaum. This excellent introduction to feedback control system design offers a theoretical approach that captures the essential issues and can be applied to a wide range of practical problems. 1992 edition. 224pp. 6 1/2 x 9 1/4. 0-486-46933-6

THE FORCES OF MATTER, Michael Faraday. These lectures by a famous inventor offer an easy-to-understand introduction to the interactions of the universe's physical forces. Six essays explore gravitation, cohesion, chemical affinity, heat, magnetism, and electricity. 1993 edition. 96pp. 5 3/8 x 8 1/2. 0-486-47482-8

DYNAMICS, Lawrence E. Goodman and William H. Warner. Beginning engineering text introduces calculus of vectors, particle motion, dynamics of particle systems and plane rigid bodies, technical applications in plane motions, and more. Exercises and answers in every chapter. 619pp. 5 3/8 x 8 1/2. 0-486-42006-X

ADAPTIVE FILTERING PREDICTION AND CONTROL, Graham C. Goodwin and Kwai Sang Sin. This unified survey focuses on linear discrete-time systems and explores natural extensions to nonlinear systems. It emphasizes discrete-time systems, summarizing theoretical and practical aspects of a large class of adaptive algorithms. 1984 edition. 560pp. 6 1/2 x 9 1/4. 0-486-46932-8

INDUCTANCE CALCULATIONS, Frederick W. Grover. This authoritative reference enables the design of virtually every type of inductor. It features a single simple formula for each type of inductor, together with tables containing essential numerical factors. 1946 edition. 304pp. 5 3/8 x 8 1/2. 0-486-47440-2

THERMODYNAMICS: Foundations and Applications, Elias P. Gyftopoulos and Gian Paolo Beretta. Designed by two MIT professors, this authoritative text discusses basic concepts and applications in detail, emphasizing generality, definitions, and logical consistency. More than 300 solved problems cover realistic energy systems and processes. 800pp. 6 1/8 x 9 1/4. 0-486-43932-1

THE FINITE ELEMENT METHOD: Linear Static and Dynamic Finite Element Analysis, Thomas J. R. Hughes. Text for students without in-depth mathematical training, this text includes a comprehensive presentation and analysis of algorithms of time-dependent phenomena plus beam, plate, and shell theories. Solution guide available upon request. 672pp. 6 1/2 x 9 1/4. 0-486-41181-8

HELICOPTER THEORY, Wayne Johnson. Monumental engineering text covers vertical flight, forward flight, performance, mathematics of rotating systems, rotary wing dynamics and aerodynamics, aeroelasticity, stability and control, stall, noise, and more. 189 illustrations. 1980 edition. 1089pp. 5 5/8 x 8 1/4. 0-486-68230-7

MATHEMATICAL HANDBOOK FOR SCIENTISTS AND ENGINEERS: Definitions, Theorems, and Formulas for Reference and Review, Granino A. Korn and Theresa M. Korn. Convenient access to information from every area of mathematics: Fourier transforms, Z transforms, linear and nonlinear programming, calculus of variations, random-process theory, special functions, combinatorial analysis, game theory, much more. 1152pp. 5 3/8 x 8 1/2. 0-486-41147-8

A HEAT TRANSFER TEXTBOOK: Fourth Edition, John H. Lienhard V and John H. Lienhard IV. This introduction to heat and mass transfer for engineering students features worked examples and end-of-chapter exercises. Worked examples and end-of-chapter exercises appear throughout the book, along with well-drawn, illuminating figures. 768pp. 7 x 9 1/4. 0-486-47931-5

BASIC ELECTRICITY, U.S. Bureau of Naval Personnel. Originally a training course; best nontechnical coverage. Topics include batteries, circuits, conductors, AC and DC, inductance and capacitance, generators, motors, transformers, amplifiers, etc. Many questions with answers. 349 illustrations. 1969 edition. 448pp. 6 1/2 x 9 1/4. 0-486-20973-3

BASIC ELECTRONICS, U.S. Bureau of Naval Personnel. Clear, well-illustrated introduction to electronic equipment covers numerous essential topics: electron tubes, semiconductors, electronic power supplies, tuned circuits, amplifiers, receivers, ranging and navigation systems, computers, antennas, more. 560 illustrations. 567pp. 6 1/2 x 9 1/4. 0-486-21076-6

BASIC WING AND AIRFOIL THEORY, Alan Pope. This self-contained treatment by a pioneer in the study of wind effects covers flow functions, airfoil construction and pressure distribution, finite and monoplane wings, and many other subjects. 1951 edition. 320pp. 5 3/8 x 8 1/2. 0-486-47188-8

SYNTHETIC FUELS, Ronald F. Probstein and R. Edwin Hicks. This unified presentation examines the methods and processes for converting coal, oil, shale, tar sands, and various forms of biomass into liquid, gaseous, and clean solid fuels. 1982 edition. 512pp. 6 1/8 x 9 1/4. 0-486-44977-7

THEORY OF ELASTIC STABILITY, Stephen P. Timoshenko and James M. Gere. Written by world-renowned authorities on mechanics, this classic ranges from theoretical explanations of 2- and 3-D stress and strain to practical applications such as torsion, bending, and thermal stress. 1961 edition. 560pp. 5 3/8 x 8 1/2. 0-486-47207-8

PRINCIPLES OF DIGITAL COMMUNICATION AND CODING, Andrew J. Viterbi and Jim K. Omura. This classic by two digital communications experts is geared toward students of communications theory and to designers of channels, links, terminals, modems, or networks used to transmit and receive digital messages. 1979 edition. 576pp. 6 1/8 x 9 1/4. 0-486-46901-8

LINEAR SYSTEM THEORY: The State Space Approach, Lotfi A. Zadeh and Charles A. Desoer. Written by two pioneers in the field, this exploration of the state space approach focuses on problems of stability and control, plus connections between this approach and classical techniques. 1963 edition. 656pp. 6 1/8 x 9 1/4. 0-486-46663-9

Mathematics–Bestsellers

HANDBOOK OF MATHEMATICAL FUNCTIONS: with Formulas, Graphs, and Mathematical Tables, Edited by Milton Abramowitz and Irene A. Stegun. A classic resource for working with special functions, standard trig, and exponential logarithmic definitions and extensions, it features 29 sets of tables, some to as high as 20 places. 1046pp. 8 x 10 1/2. 0-486-61272-4

ABSTRACT AND CONCRETE CATEGORIES: The Joy of Cats, Jiri Adamek, Horst Herrlich, and George E. Strecker. This up-to-date introductory treatment employs category theory to explore the theory of structures. Its unique approach stresses concrete categories and presents a systematic view of factorization structures. Numerous examples. 1990 edition, updated 2004. 528pp. 6 1/8 x 9 1/4. 0-486-46934-4

MATHEMATICS: Its Content, Methods and Meaning, A. D. Aleksandrov, A. N. Kolmogorov, and M. A. Lavrent'ev. Major survey offers comprehensive, coherent discussions of analytic geometry, algebra, differential equations, calculus of variations, functions of a complex variable, prime numbers, linear and non-Euclidean geometry, topology, functional analysis, more. 1963 edition. 1120pp. 5 3/8 x 8 1/2. 0-486-40916-3

INTRODUCTION TO VECTORS AND TENSORS: Second Edition--Two Volumes Bound as One, Ray M. Bowen and C.-C. Wang. Convenient single-volume compilation of two texts offers both introduction and in-depth survey. Geared toward engineering and science students rather than mathematicians, it focuses on physics and engineering applications. 1976 edition. 560pp. 6 1/2 x 9 1/4. 0-486-46914-X

AN INTRODUCTION TO ORTHOGONAL POLYNOMIALS, Theodore S. Chihara. Concise introduction covers general elementary theory, including the representation theorem and distribution functions, continued fractions and chain sequences, the recurrence formula, special functions, and some specific systems. 1978 edition. 272pp. 5 3/8 x 8 1/2. 0-486-47929-3

ADVANCED MATHEMATICS FOR ENGINEERS AND SCIENTISTS, Paul DuChateau. This primary text and supplemental reference focuses on linear algebra, calculus, and ordinary differential equations. Additional topics include partial differential equations and approximation methods. Includes solved problems. 1992 edition. 400pp. 7 1/2 x 9 1/4. 0-486-47930-7

PARTIAL DIFFERENTIAL EQUATIONS FOR SCIENTISTS AND ENGINEERS, Stanley J. Farlow. Practical text shows how to formulate and solve partial differential equations. Coverage of diffusion-type problems, hyperbolic-type problems, elliptic-type problems, numerical and approximate methods. Solution guide available upon request. 1982 edition. 414pp. 6 1/8 x 9 1/4. 0-486-67620-X

VARIATIONAL PRINCIPLES AND FREE-BOUNDARY PROBLEMS, Avner Friedman. Advanced graduate-level text examines variational methods in partial differential equations and illustrates their applications to free-boundary problems. Features detailed statements of standard theory of elliptic and parabolic operators. 1982 edition. 720pp. 6 1/8 x 9 1/4. 0-486-47853-X

LINEAR ANALYSIS AND REPRESENTATION THEORY, Steven A. Gaal. Unified treatment covers topics from the theory of operators and operator algebras on Hilbert spaces; integration and representation theory for topological groups; and the theory of Lie algebras, Lie groups, and transform groups. 1973 edition. 704pp. 6 1/8 x 9 1/4. 0-486-47851-3

Mathematics–Logic and Problem Solving

PERPLEXING PUZZLES AND TANTALIZING TEASERS, Martin Gardner. Ninety-three riddles, mazes, illusions, tricky questions, word and picture puzzles, and other challenges offer hours of entertainment for youngsters. Filled with rib-tickling drawings. Solutions. 224pp. 5 3/8 x 8 1/2. 0-486-25637-5

MY BEST MATHEMATICAL AND LOGIC PUZZLES, Martin Gardner. The noted expert selects 70 of his favorite "short" puzzles. Includes The Returning Explorer, The Mutilated Chessboard, Scrambled Box Tops, and dozens more. Complete solutions included. 96pp. 5 3/8 x 8 1/2. 0-486-28152-3

THE LADY OR THE TIGER?: and Other Logic Puzzles, Raymond M. Smullyan. Created by a renowned puzzle master, these whimsically themed challenges involve paradoxes about probability, time, and change; metapuzzles; and self-referentiality. Nineteen chapters advance in difficulty from relatively simple to highly complex. 1982 edition. 240pp. 5 3/8 x 8 1/2. 0-486-47027-X

SATAN, CANTOR AND INFINITY: Mind-Boggling Puzzles, Raymond M. Smullyan. A renowned mathematician tells stories of knights and knaves in an entertaining look at the logical precepts behind infinity, probability, time, and change. Requires a strong background in mathematics. Complete solutions. 288pp. 5 3/8 x 8 1/2. 0-486-47036-9

THE RED BOOK OF MATHEMATICAL PROBLEMS, Kenneth S. Williams and Kenneth Hardy. Handy compilation of 100 practice problems, hints and solutions indispensable for students preparing for the William Lowell Putnam and other mathematical competitions. Preface to the First Edition. Sources. 1988 edition. 192pp. 5 3/8 x 8 1/2. 0-486-69415-1

KING ARTHUR IN SEARCH OF HIS DOG AND OTHER CURIOUS PUZZLES, Raymond M. Smullyan. This fanciful, original collection for readers of all ages features arithmetic puzzles, logic problems related to crime detection, and logic and arithmetic puzzles involving King Arthur and his Dogs of the Round Table. 160pp. 5 3/8 x 8 1/2. 0-486-47435-6

UNDECIDABLE THEORIES: Studies in Logic and the Foundation of Mathematics, Alfred Tarski in collaboration with Andrzej Mostowski and Raphael M. Robinson. This well-known book by the famed logician consists of three treatises: "A General Method in Proofs of Undecidability," "Undecidability and Essential Undecidability in Mathematics," and "Undecidability of the Elementary Theory of Groups." 1953 edition. 112pp. 5 3/8 x 8 1/2. 0-486-47703-7

LOGIC FOR MATHEMATICIANS, J. Barkley Rosser. Examination of essential topics and theorems assumes no background in logic. "Undoubtedly a major addition to the literature of mathematical logic." – *Bulletin of the American Mathematical Society.* 1978 edition. 592pp. 6 1/8 x 9 1/4. 0-486-46898-4

INTRODUCTION TO PROOF IN ABSTRACT MATHEMATICS, Andrew Wohlgemuth. This undergraduate text teaches students what constitutes an acceptable proof, and it develops their ability to do proofs of routine problems as well as those requiring creative insights. 1990 edition. 384pp. 6 1/2 x 9 1/4. 0-486-47854-8

FIRST COURSE IN MATHEMATICAL LOGIC, Patrick Suppes and Shirley Hill. Rigorous introduction is simple enough in presentation and context for wide range of students. Symbolizing sentences; logical inference; truth and validity; truth tables; terms, predicates, universal quantifiers; universal specification and laws of identity; more. 288pp. 5 3/8 x 8 1/2. 0-486-42259-3

Mathematics–Algebra and Calculus

VECTOR CALCULUS, Peter Baxandall and Hans Liebeck. This introductory text offers a rigorous, comprehensive treatment. Classical theorems of vector calculus are amply illustrated with figures, worked examples, physical applications, and exercises with hints and answers. 1986 edition. 560pp. 5 3/8 x 8 1/2. 0-486-46620-5

ADVANCED CALCULUS: An Introduction to Classical Analysis, Louis Brand. A course in analysis that focuses on the functions of a real variable, this text introduces the basic concepts in their simplest setting and illustrates its teachings with numerous examples, theorems, and proofs. 1955 edition. 592pp. 5 3/8 x 8 1/2. 0-486-44548-8

ADVANCED CALCULUS, Avner Friedman. Intended for students who have already completed a one-year course in elementary calculus, this two-part treatment advances from functions of one variable to those of several variables. Solutions. 1971 edition. 432pp. 5 3/8 x 8 1/2. 0-486-45795-8

METHODS OF MATHEMATICS APPLIED TO CALCULUS, PROBABILITY, AND STATISTICS, Richard W. Hamming. This 4-part treatment begins with algebra and analytic geometry and proceeds to an exploration of the calculus of algebraic functions and transcendental functions and applications. 1985 edition. Includes 310 figures and 18 tables. 880pp. 6 1/2 x 9 1/4. 0-486-43945-3

BASIC ALGEBRA I: Second Edition, Nathan Jacobson. A classic text and standard reference for a generation, this volume covers all undergraduate algebra topics, including groups, rings, modules, Galois theory, polynomials, linear algebra, and associative algebra. 1985 edition. 528pp. 6 1/8 x 9 1/4. 0-486-47189-6

BASIC ALGEBRA II: Second Edition, Nathan Jacobson. This classic text and standard reference comprises all subjects of a first-year graduate-level course, including in-depth coverage of groups and polynomials and extensive use of categories and functors. 1989 edition. 704pp. 6 1/8 x 9 1/4. 0-486-47187-X

CALCULUS: An Intuitive and Physical Approach (Second Edition), Morris Kline. Application-oriented introduction relates the subject as closely as possible to science with explorations of the derivative; differentiation and integration of the powers of x; theorems on differentiation, antidifferentiation; the chain rule; trigonometric functions; more. Examples. 1967 edition. 960pp. 6 1/2 x 9 1/4. 0-486-40453-6

ABSTRACT ALGEBRA AND SOLUTION BY RADICALS, John E. Maxfield and Margaret W. Maxfield. Accessible advanced undergraduate-level text starts with groups, rings, fields, and polynomials and advances to Galois theory, radicals and roots of unity, and solution by radicals. Numerous examples, illustrations, exercises, appendixes. 1971 edition. 224pp. 6 1/8 x 9 1/4. 0-486-47723-1

AN INTRODUCTION TO THE THEORY OF LINEAR SPACES, Georgi E. Shilov. Translated by Richard A. Silverman. Introductory treatment offers a clear exposition of algebra, geometry, and analysis as parts of an integrated whole rather than separate subjects. Numerous examples illustrate many different fields, and problems include hints or answers. 1961 edition. 320pp. 5 3/8 x 8 1/2. 0-486-63070-6

LINEAR ALGEBRA, Georgi E. Shilov. Covers determinants, linear spaces, systems of linear equations, linear functions of a vector argument, coordinate transformations, the canonical form of the matrix of a linear operator, bilinear and quadratic forms, and more. 387pp. 5 3/8 x 8 1/2. 0-486-63518-X